T0122345

Time Series Analysis Methods and Applications for Flight Data

Jianye Zhang · Peng Zhang

Time Series Analysis Methods and Applications for Flight Data

Jianye Zhang
Air Force Engineering University
Xi'an, Shaanxi
China

Peng Zhang
Air Force Engineering University
Xi'an, Shaanxi
China

ISBN 978-3-662-57136-1 ISBN 978-3-662-53430-4 (eBook)
DOI 10.1007/978-3-662-53430-4

Translated from Chinese
Jointly published with National Defense Industry Press, Beijing, China

Printed on acid-free paper

This Springer imprint is published by Springer Nature
The registered company is Springer-Verlag GmbH Germany
The registered company address is: Heidelberger Platz 3, 14197 Berlin, Germany

Preface

Flight data automatically recorded and saved by FDRS in aircraft flight and maintenance is a typical type of time series. The flight data of this type is a recording of the operation states of the aircraft and its subsystems, as well as the change trends of aircraft performance. More importantly, the accumulation of the data along with time will produce massive data that can be used as a source of authentic onboard evidence of aircraft performance and condition. Understandably, it can also be used as a predicator of operation condition of aircraft systems for the early warning of a possible malfunction. Therefore, effective use of the flight data, an area that has recently attracted much research effort, provides a reliable basis for on-condition maintenance, quality control, flight safety assessment, and air accident investigation as well.

While the development of science and technology benefits the world in the information age, there is also an increasing conflict between data redundancy and lack of knowledge. Information mining out of massive data on the one hand, and transformation of the mined information into applicable knowledge on the other, poses a challenge to the research and application of flight data. As a result of this, research efforts have been made to answer to the challenge, and data mining is one of them. As part of the effort to explore the theory and practice of data mining in flight data analysis, and on the basis of the view of flight data as typical time series, the present study provides a detailed and systematic account of the aim of flight data analysis, as well as its procedures and implementation technologies. Integrating in it the latest development in control theory, test technology, and information processing, the present study is also a summary of the contribution of the research team to the study and application of flight data in the past 10 years.

The present volume is made up of six chapters. Chapter 1 is an introduction of the research background, recalling and summarizing the basic notions, developments, and military application of the study of flight data. Chapter 2 deals with flight data preprocessing, providing a theoretical basis, as well as a reliable data source, for the discussion and practice of intelligent data mining in the following chapters. Chapter 3 is an introduction of ARMA model for time series analysis, and its application and technological realization. In Chap. 4, methods of similarity search of time series are proposed together with a theoretical account of the methods. Chapter 5 focuses on elaborating flight-data-based second development and advanced application for the purpose of condition monitoring, failure diagnosis, and trend prediction of aircraft and its subsystems by means of various data mining methods. Chapter 6 introduces system design and application of intelligent flight data mining.

Many people have contributed to the completion of this book; among them are Profs. Li Xue-ren and Ni Shi-hong from Air Force Engineering University (AFEU), and Prof. Pan Quan from Northwestern Polytechnical University (NWPU). We gratefully acknowledge their help and guidance. It should also be acknowledged that Liang Jian-hai, in his a postdoctoral research at NWPU, has contributed to search and prediction algorithms. Our thanks also go to Song Ji-xue, Liu Bo-ning, Wang Zhan-lei, postgraduates at AFEU, who have helped with the writing and proofreading of some of the draft. Last but not least, we are grateful to all the authors who have been consulted and quoted in the process of writing.

The authors hold themselves to be responsible for all the possible errors in the volume, and welcome suggestions for its improvement.

The book has been a joint effort of six translators who are all English teachers at AFEU. They are Zhao Donglin (Chap. 4), Wu Subin (Chap. 1 and 2), Zhang Jing (Chap. 5), Wang Ningwu (Chap. 5), Jiang Fei (Chap. 3), and Jiang Xuehong (Chap.6). The completion of the English version is also a result of constant consultation and discussion with the authors for appropriate comprehension and re-rendering of the original text. The whole translation has been reviewed and revised by the authors, Zhao Donglin and Wu Subin.

Xi'an, China
August, 2016

Jianye Zhang
Peng Zhang

Contents

Chapter 1
Introduction

This chapter provides background information for the contents of this book. It briefly introduces some basic concepts about Flight Data Recorder System (FDRS) and its developments. Meanwhile, it focuses on the status and development trends of flight data application research in assessing flight quality and flight performance, monitoring off-line/on-line equipment status and investigating accidents. Then, the research area and main contents of this book are presented with focus on the flight data as a typical time series.

1.1 Flight Data Recorder System

1.1.1 Overview

FDRS is designed to collect and record data concerning the performance of an aircraft and its subsystems. It provides an objective and scientific reference for fault diagnosis of aircraft, aiding flight training, and flight accident investigation. In this book, the data concerning the performance of an aircraft and its subsystems collected and recorded by the FDRS is also named flight data.

Nowadays, generally speaking, FDRS consists of Input Data Source, Flight Data Acquisition Unit (FDAU), Flight Data Recorder (FDR) and Quick Access Recorder (QAR). Figure 1.1 shows the FDRS framework.

According to Chinese military specifications, FDR is defined as "an airborne automatic recorder which records data about flight status, control status as well as aircraft/helicopter and engine." The primary function of the airborne flight data recorder is to accurately record and effectively save data about various flight statuses. Performance indexes mainly include recording capacity, bit error rate, and crashworthiness. Recording capacity consists of such indexes as the number of data simultaneously recorded, recording time, and recording process. Bit error rate

J. Zhang and P. Zhang, *Time Series Analysis Methods and Applications for Flight Data*, DOI 10.1007/978-3-662-53430-4_1

Fig. 1.1 FDRS framework

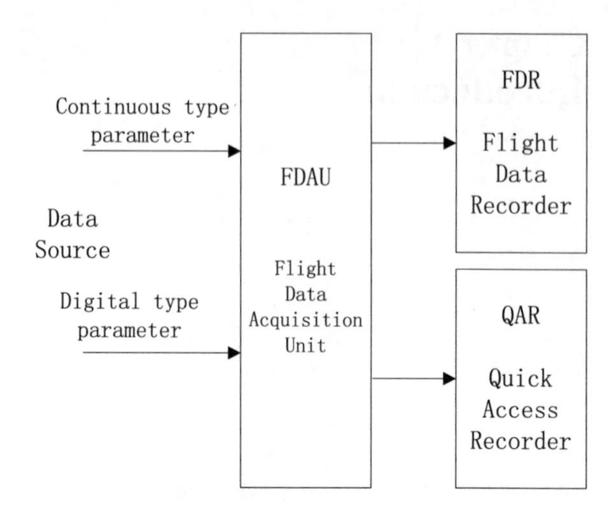

Fig. 1.2 Outward appearance of a crash-protected data recorder

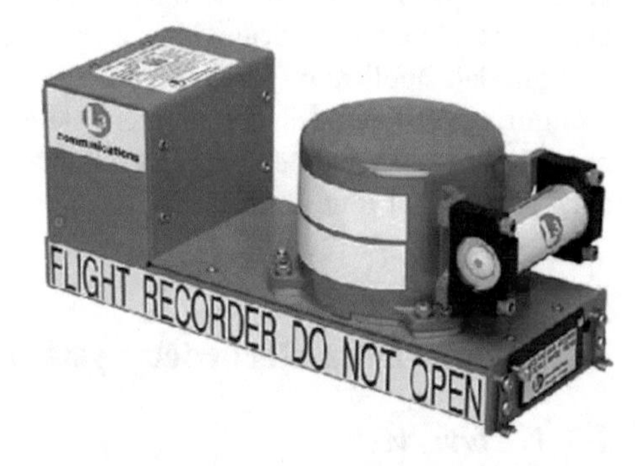

indicates whether the FDR can record normally. Crashworthiness of the recorder includes such performance data as bearable gravity acceleration, heat resistance, pressure resistance, and acid resistance.

Depending on whether it has crashworthiness, FDR can be divided into two types: maintenance FDR without crashworthiness and anti-incident-destruction FDR with crashworthiness. The latter is commonly called "black box" which is the carrier of objective evidence that is indispensible for flight accident investigation, as shown in Fig. 1.2.

1.1.2 Developments

1.1.2.1 FDR in the Early Days

The history of FDR can be dated back to the beginning of the electricity-powered flight age. The first flight in human history by Wright Brothers was "eternally

Fig. 1.3 FDR equipped in "Louis Spirit"

recorded" by the first FDR. The original device recorded only a few data such as rotor speed of the propeller, flying distance, and duration in the air. Charles Lindbergh, another flying pioneer, equipped self-designed "Louis Spirit" with a more advanced flight recorder. The recorder was installed in an index card-sized wooden box, as shown in Fig. 1.3. The automatic barometer inside the recorder could record the changes of air pressure or altitude in a rotating paper column. Regrettably, these two FDRs did not survive an accident due to lack of crashworthiness.

1.1.2.2 Nick-Typed FDR

While commercial aviation was flourishing in the 1940s, frequently occurring flight accidents triggered increasing concern from the Civil Aeronautics Board (CAB) about the importance of flight data. The CAB worked with various companies to explore more reliable ways to record flight data.

To meet this challenge, General Electrics developed "selsyns" system in which a series of mini-electrodes directly installed on the aircraft equipment serves as a sensor and transmit information to the recorder in the rear of the aircraft through cable. During the design process, engineers of General Electric Company overcame a series of technological challenges, for example, to prevent possible icing or probing pen blocking by the ink of the recorder at high altitude with low air pressure and low temperature, the engineers used a probe to cut a figure on a black sheet covered with white paint. But regrettably, this recorder was not used in actual flight.

Meanwhile, Frederick Flader engineering company developed a tape recorder, but this recorder was still not used in flight. "Black box" technology did not make great progress until 1951 when James J. Ryan joined the machine department of the General Mills Company. Considering the problems with FDR, Ryan designed a "VGA flight recorder" to meet the engineering application requirement. In VGA,

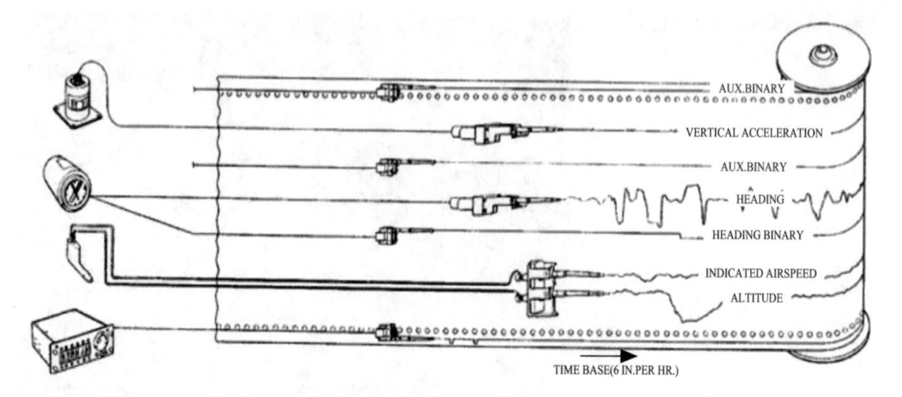

Fig. 1.4 Functional diagram for the typical platinum bar nick-typed FDR

V stands for velocity, G for gravitational acceleration, and A for altitude. VGA flight recorder was about 10 lb with two separate cases, one for measuring device, the other for a recorder connected with the measuring device. It is noteworthy that nowadays this kind of design is still widely adopted in FDRS.

The first generation crash-protected FDR in real sense was not produced until 1953. The recording media mainly included steel belts or films. The recorded data were handled only when a major incident happened or an analysis was actually required. A typical platinum bar nick-typed FDR used a firing pin to line out "oscillographic display" corresponding to every parameter on the platinum bar, as shown in Fig. 1.4. Accident investigators read out these "nicks" through a microscope, then decoded the deviation of these nicks from the reference lines into engineering values. The whole process was complicated and time consuming.

In 1957, the Civil Aviation Authority(CAA) published the third edition flight recorder standard, stipulating that by July 1, 1958, all transporters heavier than 12,500 lb with flying altitude over 25,000 ft must be equipped with an anti-destruction FDR in compliance with TSO C-51 technical standard. This standard defines some special requirements for flight data in such aspects as accuracy, sampling intervals, parameter types(altitude, airspeed, heading, vertical acceleration and time), crash-protected capability, and capability of surviving up to 30-min burning at 1100 °C. The imperative implementation of this standard opened a wide commercial market for FDR.

Regrettably, most recorders initially were installed near the cockpit or the main airplane wheel where the recorders were very likely burned or suffered from heavy pressure, thus being destroyed or severely damaged. In the 1960s, CAB put forward a suggestion to the Federal Aviation Administration (FAA): FDR should have additional protection function against heavy pressure and burning and FDR should be moved to the tail of the aircraft to gain the maximum protection to the recorder. Therefore, FAA revised regulations, stating that the recorder should be installed as close to the tail of the aircraft as possible; meanwhile FAA upgraded the performance standard from TSO C-51 to TSO C-51a, as shown in Table 1.1. But neither

Table 1.1 Comparisons between TSO C-51 and TSO C-51a

	TSO C-51	TSO C-51a
Fire resistance	50% capable of surviving 30-min burning at 1100 °C	50% capable of surviving 30-min burning at 1100 °C
Impact vibration	100 Gs	1000 Gs
Extrusion	No requirement	Capable of bearing 5000 lb within 5 min in monaxial direction
Liquid soaking	No requirement	Can be soaked in aviation liquid (fuel, oil, etc.) within 24 h
Seawater soaking	Can be soaked in seawater for 36 h	Can be soaked in seawater for 30 days
Penetrability	No requirement	A 500-pound object with 1/4-in. diameter interface falling from 10 ft above should not penetrate the recorder

of the two standards provided a proper test agreement to guarantee common repeatable test conditions.

1.1.2.3 Tape-Typed FDR

On October 10, 1972, Digital Flight Data Recorder (DFDR) that could record more flight data was considered as the new standard, thus enabling tapes to become the preferred recording medium. That situation continued until the solid FDR appeared in the late 1980s. At that time, FDR manufacturers used a variety of tapes and tape units. Figure 1.5 shows the F800 DFDR tape unit and its protective shell.

Mylar tapes, polyimide tapes, and metal tapes were the most widely used tapes at that time. The types of tape units included coplanar rotation, coaxial drive, looping

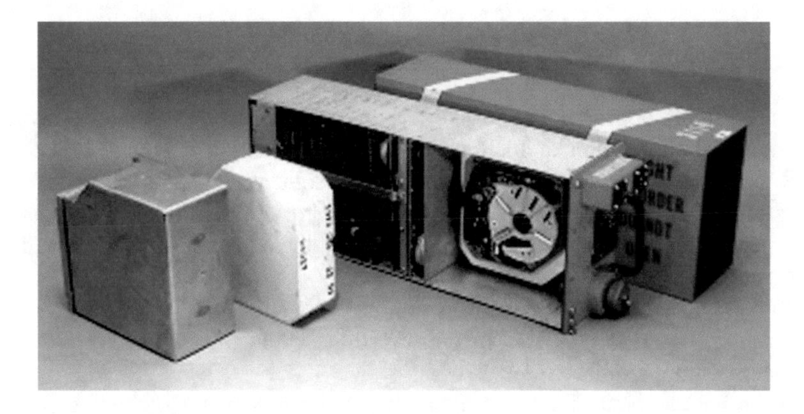

Fig. 1.5 F800 DFDR and Its protective shell

wiring, random access memory, etc. However, the lethal shortcoming of all tape FDRs (including metal tape units) was that they could hardly survive the fire after crash. The high-intensity fire resistance test was needed, but it was still difficult to adopt an appropriate design for lack of detailed test standards. Furthermore, experience shows that metal tapes are vulnerable to pressure shock because they could make springs lose tension, thus leading to severe damage of the tapes and loss of data.

It is worth mentioning that a cockpit voice recording function was added into recorders, i.e., Cockpit Voice Recorder (CVR), providing "audio" evidence for flight accident investigations. FDR and CVR usually adopt recurrent recording mode.

Although QAR and DFDR appeared almost at the same time, the former belonged to a kind of optional recorder. They both adopted same recording technology, but QAR could store and unload flight data more quickly, and could record more types of parameter. Most DFDR and QAR needed a FDAU to provide an interface between sensors and QAR/DFDR. FDAU could transfer analog signals from sensors into digital signals, and then combine them into a serial data flow for recording by means of multiplexing technology. For tape recorders, the data flow format in industry is 64 12-bit data words per second. Because recording capacity was limited by tape length, although the capacity was enough for the first generation wide-body transport aircraft, it became insufficient after the appearance of aircraft equipped with digital electronics, such as Boeing 767 and Airbus A320.

In the early 1980s, Digital Avionics Systems (DAS) were introduced into the commercial aviation, increasing dramatically the amount of information from DFDR and QAR. DAS applied digital data bus technology that could transmit digital data among systems, which made it easier to obtain a mass of important flight data and aircraft system information from the data bus. Furthermore, the advent of data bus advanced development of digital FDAU module (DFDAU). DFDAU executed the same function as FDAU did, whereas DFDAU could provide the interface between the data bus and the analog sensor.

1.1.2.4 Solid Status FDR

The Solid FDR appeared in the late 1980s, which marked the most important advance in the evolution of FDR technologies. The shape of typical DFDR and CVR recorders is shown in Fig. 1.6. Adoption of solid storing device brought about two significant advantages: first, it expanded recording capacity of flight data and improved usability of the data; second, it strengthened crashworthiness and fire resistance and improved recording reliability.

With concerted efforts of accident investigators and Recorder Industry Association, the new standards of crashworthiness and fire resistance have been released, as shown in Table 1.2.

At the International Flight Data Recording System Annual Conference in 1998, some requirements were proposed as follows: ① The aircraft manufactured after

Fig. 1.6 Shape of typical DRDR and CVR

Table 1.2 Standards of crashworthiness and fire resistance of present flight recorders

	TSO C-124a (digital flight data recorder) and TSO C-123a (cockpit voice recorder)
Fire resistance (high intensity)	Capable of surviving 30-min burning at 1100 °C (capable of surviving 60-min burning with ED56 test agreement)
Fire resistance (low intensity)	Capable of surviving 10-h burning at 260 °C
Impact vibration	3400 g within 6.5 ms
Extrusion	Capable of bearing 5000 lb within 5 min in monaxial direction
Liquid soaking	Can be soaked in aviation liquid (fuel, oil, etc.) within 24 h
Penetrability	A 500-pound object with 1/4-in. diameter interface falling from 10 ft above should not penetrate the recorder
Static liquid pressure	The pressure equivalent thickness is 20,000 ft

August 18, 2002 needs to increase the number of data recorder parameters from 57 to 88, with specific requirements for the parameters. ② The aircraft manufactured after January 1, 2003 needs to combine the previous crash-protected recorder and maintenance recorder into a consolidated recorder. Each aircraft will be equipped with two such recorders for the sake of reliability. ③ After 2005, the crash-protected recorder on board should be transformed into the solid state recorder, together with one additional cockpit image recorder. The solid state storage units make it possible to install a cockpit video recorder. In this way, the actually shown images can be connected with the objective records of the flight data through recording the video images of the main instrument panels.

With wide application of computer technologies and data buses on the aircraft, flight data collected and recorded by the FDRS increases significantly. The functions of the FDRS are not only to understand test flights and accident investigations, but to provide the ability to monitor aircraft status and pilot's operations.

1.1.2.5 Advanced Digital Data Recording System

In order to avoid the repeated changes of recorders and their interface for other systems, it is urgent to develop a plan for advanced digital data recording system. This new architecture views current extended data acquisition system as a central process unit, thus various high-speed serial interfaces can interchange with dual Solid State Digital Data Recorders (SSDDR), as shown in Fig. 1.7.

This architecture inherits current crash-protected recording system to a great extent. In the improved system, the crash-protected recorder is converted into a recorder for digital information from high-speed serial interfaces. There is no need for the recorder to know the defined information sources and types, except recording the digital data under the predefined rules. The information processing and digitalization process can be done in other airborne systems, and then sent to redundancy recorders. In this architecture, the crash-protected recorder needs enough storing capacity and input bandwidth to meet current and future requirements. Furthermore, SSDDR partitions can distinguish data types, thus enabling optional downloading of flight data. The advantages of this architecture include:

1. Dual redundancy crash-protected recorder can reduce the layout of aviation cables, and also greatly improve data recovery rate;
2. It can improve processing efficiency. The procedures of data processing and discretization can be finished in the highly integrated avionics;
3. It can reduce cost. The crash-protected recorder can meet the change requirement without any correction, thus diminishing the requirement for installing another crash-protected recorder to store additional information.

1.1.2.6 Enhanced Airborne Flight Recorder

In a technology demonstration in 2007, the FAA demonstrated the product of Enhanced Airborne Flight Recorder (EAFR) to the International Civil Aviation

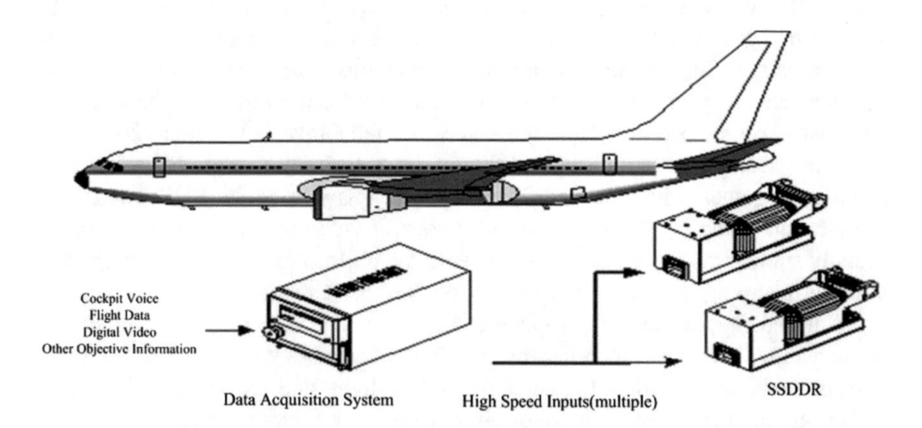

Fig. 1.7 Framework of advanced digital data recording system

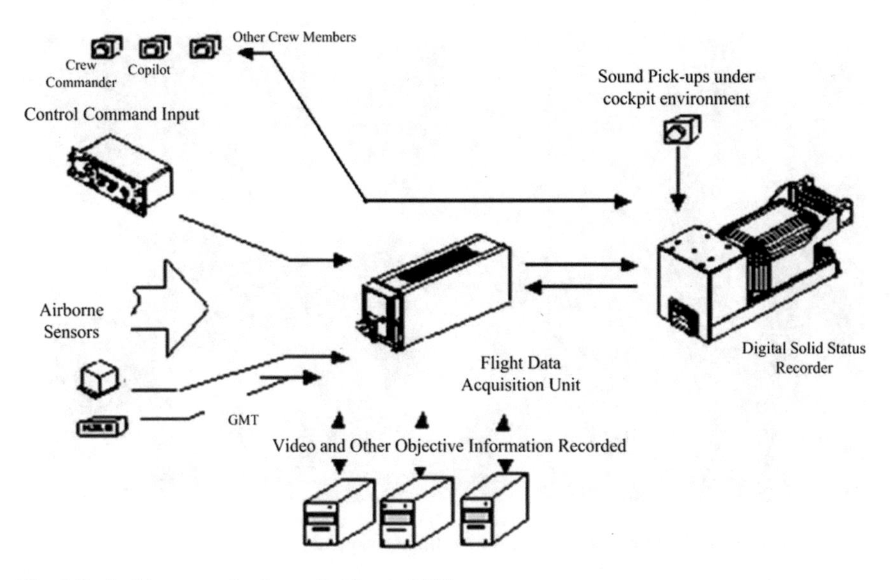

Fig. 1.8 Architecture of enhanced airborne FDR system

Organization (ICAO). This marked materialization of the global standards for FDRS developed concertedly for 10 years by aviation industry and U.S. government departments. As shown in Fig. 1.8, the architecture of EAFR mainly includes the following functional modules: digital solid FDR, CVR, data-link data records, video records, and comprehensive acquisition of flight data. These functional modules designed as a more reliable Line Replaceable Unit (LRU) take the place of the previous independent hardware.

Boeing 787 aircraft owns an avionic network architecture. A set of EAFRS produced by U.S. General Electric Company is installed in the nose and tail of the aircraft respectively, as shown in Fig. 1.9. With current configuration, each EAFR includes the functions of CVR and FDR, resulting in system redundancy; furthermore, it can provide enough capacity for recording 2-h video information. This EAFRS can record nearly 2000 parameters with 50-h recording time; while the standard of current airborne recorders needs at least 88 parameters, with 25-h recording time. Compared with the traditional flight data system, the separate flight data acquisition unit is no longer an "actual FDAU". Each EAFR is put in other software and hardware including EAFR, and can receive flight information data directly from aircraft sensors and other subsystems. This design will significantly reduce the overall weight of systems.

With the development of aviation industry, FDRS has experienced the course from its birth to development, as a result the overall performance has been improved and functions expanded. Throughout the hundred-year development history of flight data systems, the previous recorders such as nick-typed FDR, wire-typed FDR, and tape-typed FDR have been replaced by solid storing FDR; the number of acquisition signals has increased from several to dozens, even hundreds

Fig. 1.9 Boeing 787 and its enhanced airborne flight recorders

and thousands; the frequency of signal acquisition data has improved from several times per minute to several times per second; the mode of signal processing and analysis has developed from manual to automatic processing; the information exploitation has been expanded to such areas as accident investigations, aircraft status monitoring, and flight operation assessment. All of these play an important role in improving flight quality and ensuring flight safety.

1.1.3 Military Applications

In America, military flight data system evolved into the Flight Data Management System (FDMS) in the 1970s. And it is an important technology to develop advanced fighters and other types of military aircraft. FDMS is no longer a pure system for data acquisition and information recording, but a comprehensive processing and analysis system consisting of three parts (airborne, line maintenance and ground service). It can not only collect and record the whole flight data information in real time, but can transform it into useful information immediately. FDMS has obtained high technological competence for the comprehensive process and analysis of flight data information. It not only has the processing capability with general interface of various airborne equipments, but also has the capability to store large amount of data and transmit them at high speed in real time. The flight data information managed by FDMS covers information of airborne systems such as engine system, flight control system, air data system, integrated cockpit display and control system, fuel measurement system, fire control system, radio communications system, navigation, and power supply system. At present, FDMS has been installed on some fighters such as F-15, F-16, Mirage 2000 and F-22.

1.2 Application Research for Flight Data

1.2.1 Basic Concepts

The airborne flight data system is a "measuring-sampling-recording" device in nature, which only records the information sampled at the preset time. Therefore, according to their numerical characteristics, the flight data can be categorized into continuous parameter type and discrete parameter type, as shown in Fig. 1.10. The figure shows that flight data is a kind of typical time series data. The ordinate in the display area of the flight data of continuous type shows the numerical value of current parameter. The protuberant part in the display area of the flight data of discrete type shows that the status of current parameter is "1". Otherwise the status is "0". The abscissa shows the relative time (unit: second) when the flight data are recorded.

The flight data of continuous parameter type are from the airborne sensors or data bus, which are the data of aircraft status attributes shown as real number at discrete sampling time, such as flight altitude, engine rotating speed, longitude, and latitude. Different from flight data of continuous type, the intermittence of flight data of discrete type is also displayed on the amplitude, which is the nondimensional data of aircraft status attributes shown as numerical status value "1" or "0" at discrete sampling time, such as gear up or down status, flaps up or down status, etc.

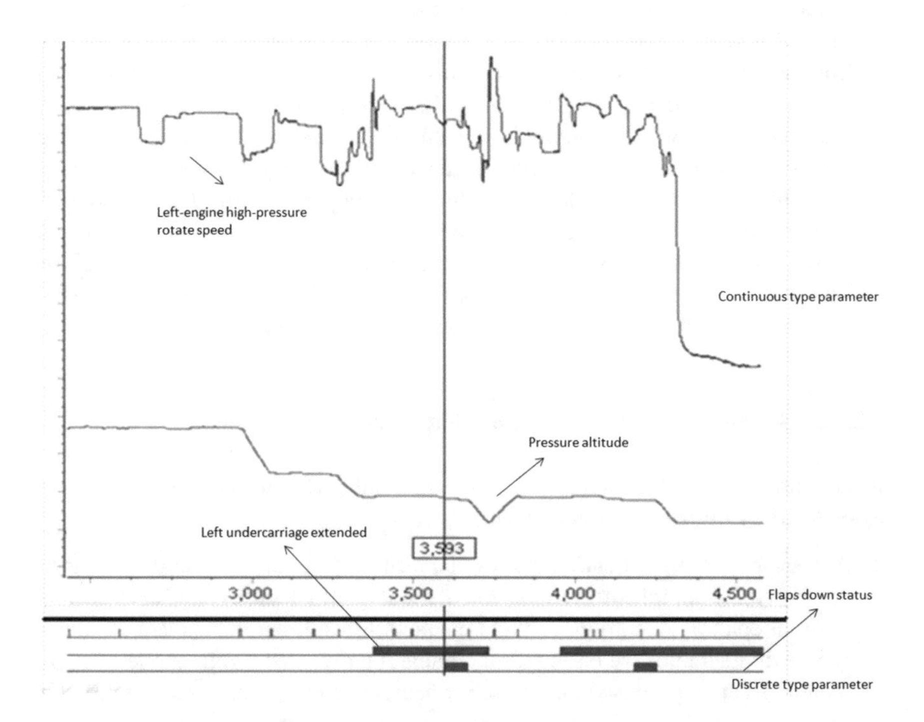

Fig. 1.10 Diagram of continuous type and discrete type flight data

The flight data of discrete type depict the status or attribute of the aircraft or its subsystem at corresponding time. It may also be a specific event.

1.2.2 Research Status

According to the differences between flight data handling process and application, the study based on further development and intelligence handling of flight data mainly includes preprocessing of flight data, assessment of flight quality and performance, off-line and on-line equipment status monitoring, accident investigation, etc.

1.2.2.1 Preprocessing of Flight Data

Practice shows that producing high-quality flight data is important and fundamental, and is also necessary for subsequent intelligent processing of flight data. The preprocessing process of flight data is to conduct the process of analyzing the raw data recorded and eliminating false data and saving true ones.

Some traditional ways, such as Chauvenet's Criterion Test, arithmetic average filtering algorithm, weighted moving average filtering algorithm and inertia filtering algorithm, can realize elimination of outlier data in the flight data to some degree. The neural network based on artificial intelligence and support vector machine can realize prediction and assessment of the missing value in flight data by using their own particular nonlinear mapping ability. However, when parameters change, the remodeling process would be complicated and time consuming. In the international engineering community, it is universally acknowledged that before pattern recognition, feature selection should be conducted. Dengfeng Zhang has realized the feature extraction of raw samples by adopting rough set theory. James H. Taylor discussed the chaos theory in the nonlinear dynamic system. He analyzed the chaotic property by using real-time flight data from helicopters, especially the two types of accelerated velocity data series with sampling frequency of 1024 Hz.

1.2.2.2 Assessment of Flight Quality and Aircraft Performance

It is important to assess flying abilities of a pilot and to conduct aircraft performance assessment by utilizing recorded data.

1. Assessment of flight quality. For military aircraft, because of its great maneuverability and complexity, assessment of flight quality focuses on standardization of flight control. According to flight characteristics, flight maneuvers of aircraft can be divided into five basic types: circulating, dive, loop, roll, and level flight. Yufeng Li et al. proposed that for fuzzy neural network, supervised learning law be adopted to identify basic flight patterns and standard flight maneuvers, which

realized clustering analysis and effective segmentation of data. The authors of this book, based on manual interpretation experience, have researched and developed the Flight Training Automatic Evaluation and Management System, which can automatically identify flight maneuvers, and generate a flight maneuver list with related flight maneuver data. The flight operation of civil aircraft is comparatively simple, and its flight quality assessment mainly focuses on safety. Takeoff, landing, and en route flying are three basic statuses of the whole flight process, but takeoff and landing are the flight phases when flight accidents are prone to happen. In the process of processing flight data, considering the basic operating status of aircraft and many emergency situations (for example, go around), the whole process is divided into five phases: taxi out and takeoff, liftoff and climb, fly en route, prepare for landing, and land and taxi back. Then, it is necessary to compare it with standard mathematical model or with database model to accomplish index evaluation. The standard model proposed by Yongfang Huang was established upon the flight manual and quality standard for a pilot's solo, and the flight grades were judged according to the flight accident module and pilot skill evaluation module. Jianhai Liang proposed that imprecise reasoning be adopted to set mark positions flexibly, and this classification method could classify the flight phases effectively.

2. Assessment of aircraft performance. Assessment of aircraft performance by using flight data mainly focuses on such aspects as aerodynamics parameter estimation and functional assessment of flight control system. Yoichi conducted an adaptive estimation of the aircraft stability derivative by adopting the method of inputting flight data into composite observer. Giampiero introduced time domain-based parameter identification (PID) method to estimate the aerodynamic coefficients of a specific aircraft. The basic method is the Local Weighted Regression (LWR) technology. The resulting PID algorithm was tested using NASA F/A-18 flight data to evaluate the real-time capabilities of on-line PID. The results were then compared with theoretical estimation from wind tunnel analysis and with the PID results. Bennett, by adopting FDR and QAR flight data, studied the risk assessment model of aircraft performance. After marking incident signs and accident risk, the unusual information of the FDR was input into the hybrid modeling system, and effective output was produced through expert system rules.

1.2.2.3 Monitoring of Off-Line and On-Line Equipment Status

In order to monitor the aircraft equipment status, the flight data of all types of aircraft, in most cases, have a corresponding expert system. One characteristic of expert system is to effectively use the abundant knowledge and experience of experts in relevant fields to make up for lack of knowledge acquisition. The knowledge reasoning machine usually represents expert knowledge through production rule.

1. Off-line monitoring. Currently, most fight data recorded concern the engine. Take one type of transporter as an example, of about 100 parameters recorded, 28 signals of 7 types are about engine, which account for 30% of all the recorded parameters, including rotor speed, position of the throttle lever, exhaust temperature, position of shutdown handle, switch-on thrust reverser, and dangerous vibration. In integrative parameter analysis, Yingjun Wang et al. studied on the role of flight data in monitoring aero-engine performance tendency. They proposed that integrative parameters be used as a quantitative criterion for judging engine performance. In single parameter analysis, Yubo Wang et al. conducted modeling and analysis of the data recorded in flight data system by time series analysis theory. They chose gas temperature of the engine to form unary time series, and proposed one method to monitor aircraft engine status. Dennice used flight data for fault detection and diagnosis in gas turbine engines. The fuzzy logic rule base is derived using heuristics based on designed experiments and flight data. The method is evaluated using model-based residuals and calculated values as inputs. They also studied the method of augmenting input parameters to better the prediction horizon for diagnosis. It is proved that this method is effective in fault detection of high pressure spool deterioration of engines. Vaughn, by using Multilayer Perceptron, predicted helicopter airframe load spectra from continuously valued flight data for two example regression outputs—a high damage case and a low damage case. Marcello realized reconstruction of aircraft control signals by suing neural network and fuzzy logic. They can be used in the investigation of an uncommanded maneuver, and can also simulate control surface deflections. This method has been applied effectively in B737-300.
2. On-line monitoring. On-line (real time) monitoring refers to collecting data from FDR or other types of flight data, continuously and accurately transmitting all types of real-time flight data to the ground by using data communication and computer simulation and graphics/image processing technology. Thus, the commander knows the aircraft status, including the equipment status and operating information, checks the flight condition, and conducts effective command. Chunsheng Xu conducted a research of application of real-time monitoring technique in civil aircraft maintenance, and established aircraft communication addressing and reporting system. This system needs to upload and download information data and directive information simultaneously. Before landing, the maintenance suggestions are provided by the maintenance software to maintainers. The authors of this book have studied such issues as communication between the ground station and airborne equipment, real-time data receiving and processing methods by the ground station, and data transmitting and receiving key technical issues between computer networks in the flight real-time monitoring system.

1.2.2.4 Accident Investigation

Data being monitored during flight come from diverse systems of the aircraft. Besides the flight data system, other objective information sources include cockpit voice recorder and integrated built-in test and fault warning system, and fire control information recording system for a military aircraft. Information fusion method can be utilized to conduct integrated analysis of the different information. The observed information acquired in a time series manner is automatically analyzed, integrated, and applied under certain rules to get concordant interpretation and description of the object being detected so as to achieve required decision-making and estimation. Shihong Ni et al., by using one type of aircraft, analyzed different types of sources upon which flight accident investigations are dependent and the operating characteristics of the sources, discussed time-based synchronization of various information sources, and provided the time-based synchronization method.

1.2.2.5 Other Applications

According to the flight data recorded by FDR, flight simulation from the angle of the pilot or ground observers can be produced by computer simulation technologies, which can vividly reproduce the flight status of various situations and provide visual analysis method for accident analysis, routine maintenance, and flight quality monitoring. Current applications of simulation technologies focus on study and construction of semi-physical simulation composed of real flight data and pilots, fly-by-wire system, and flight control system to reproduce and verify the actual flight status in the air. Yuehua Yu et al. proposed the concept of engineer flight simulation system. According to the requirements of specified flight simulators, and based on real-time monitoring system, this system is designed to have modules for multi-threading management, monitoring curve drawing and simulation data handling, which can conduct pre-flight simulation test for airborne systems. Tong Mei et al. introduced a method for texture collection and handling in flight scenes, flight data handling, modal structure, and software programming for depicting flight process.

Andrew proposed the concept of Flight Object (FO). FO extracts actual data from all fight data sources. The data include aircraft type, estimated data, trajectory, equipment, weight, and so on. FO has already passed through the initial feasibility phase and is now entering the development phase. As new interoperability standard to be used in Europe for the specifications of new flight data processing systems, it has been deployed for actual application since 2007. At the same time, it has been proposed to the ICAO for global standardization.

Levine proposed the concept of remote aircraft flight recording and advisory system with the objective of establishing a global aircraft data expressway. This system integrates the data from DFDR sensors and those from ATC, including GPS/GLONASS data, map, terrain, weather information, and so on. The data are encrypted, transmitted, and shared for accident prevention. They can also be used to estimate the position of crashed aircraft, thus making search and rescue easier.

Sebastian designed flight parameter handling software based on data reconstruction. He used “software black box” to reconstruct the data for reproduction of fault mode. Shuli Gong described the flight parameter data unloading and data management system from the perspective of hardware structure, which integrated ARINC717 data collection unit and aircraft environment monitoring function. Kam adopted acoustic Lloyd’s mirror effect to verify effectiveness of flight data. The specific plan is to install a microphone above the ground where the aircraft would fly over to estimate the aircraft position parameters and aircraft motion parameters.

Euro Maastricht Upper Area Control Centre has already installed the next-generation interoperable FDRS, providing air traffic control for traffic over Belgium, Luxembourg, Netherlands, and northwest Germany, one of the busiest and most complicated airspace in the world. This system was manufactured by Indra, a Spanish system provider. It supports both civil and military operation, and also provides a series of state-of-the-art tools. It can handle the ever-growing air traffic flow by using a safer, more convenient and environment-friendly method. The design of this system conforms to the long-term goal of Single European Sky Air Traffic Management Research Plan and European Infrastructure Modern Plan P4YJ, and will be helpful for realizing the objective of a more integrated performance-driven Air Traffic Management system. This system can manage airspace more flexibly, for example, it can conveniently redivide the airspace, dynamically respond to the change of real-time traffic requirements, special weather and set aside airspace for military operations. Meanwhile, good adaptability of the system will guarantee the low rate of flight delays, thus improving the overall operational performance of ATC.

1.3 Research Domain and Main Contents

Flight data is a kind of objectively recorded information to describe the flight process of an aircraft. It has three-aspect attributes of time, space, and condition. The information structure based on time, space, and condition, which appears to be three-dimension super space, forms the applied framework of information source development concerning the mining of flight data series information.

As to the intelligent mining and process of flight data with relation to the time-dimension attribute, the focus is to make use of the massive amount of flight data to search for useful information for flight guidance and maintenance assistance. As to the intelligent mining and process of flight data with relation to the space-dimension attribute, the focus is to make use of the space data concerning flight attitude status, such as flight altitude, heading angle, angle of pitch, and flight velocity to realize automatic identification of flight trajectory, and carry out flight quality evaluation and accident analysis. As to the intelligent mining and process of flight data with relation to the condition-dimension attribute, the focus is to make use of status data which indicate “health conditions” of the subsystems or airborne equipment of the aircraft, such as exhaust gas temperature and engine speed to conduct monitoring, diagnosis and trend prediction.

The focus of this book is on the research of flight data with relation to the dimensions of time and space, with limited coverage of the space attribute of flight data.

Chapter 2
Preprocessing of Flight Data

Research findings indicate that the amount, accuracy, and type of data collected from airborne FDRS can hardly meet the requirements for further application and development of flight data. Therefore, it is necessary and fundamental to preprocess original flight data. The level of accuracy and reliability of preprocessing results will have a direct bearing on the quality of follow-on research. This chapter covers outlier elimination, data filling, data extension, reduction of monitorable parameters and chaotic property analysis.

2.1 Support Degree-Based Amnesic Fusion Filtering Method for Flight Data

Diverse and redundant data collected and recorded from FDRS are information sources for application of data fusion filtering technologies. In this section, amnesic fusion filtering method of flight data based on support degree is introduced, which greatly supplements and expands current outlier preprocessing algorithms. Exponential functions are used to build a matrix of support degree to conduct weighting fusion of measurement data, thus eliminating the dependence of the algorithm upon priori knowledge. Meanwhile, an amnesic control item is added to dynamically adjust the fusion weights according to data, thus preventing the "data saturation" caused by excessive data.

2.1.1 Unified Error Model of Flight Data

Every measurement result has errors. Errors exist in every scientific experiment and measurement. Webster defines error as "the difference between an observed value

J. Zhang and P. Zhang, *Time Series Analysis Methods and Applications for Flight Data*, DOI 10.1007/978-3-662-53430-4_2

and the true value of a quantity." There is no exception for FDRS. As a typical real-time data measuring and recording system, FDRS is subject to a variety of interferences during its operation. Thus, outliers are inevitable in the flight data. These abnormal data will have a direct influence upon follow-on processing. Only through sorting out the quality of these flight data can be guaranteed.

Measurement errors are determined by such factors as measuring environment, times and moment. Measurement errors fall into three categories: system errors, random errors, and gross errors. System errors can be reduced by improving the performance of equipment while random errors and gross errors can be reduced with application of appropriate mathematic methods.

As for raw flight data, the relationship between the true value and the measurement value is as follows:

$$X'(k) = X(k) + \Delta e0 + \Delta e1 + \Delta e2 + \Delta e3 + \Delta e4 \quad (2.1)$$

In the formula, $X'(k)$ stands for the measurement value of parameter X at sampling time k; $X(k)$ for the true value of parameter X at sampling time k; $\Delta e0$ for random interference error; $\Delta e1$ for sensing error of the sensor; $\Delta e2$ for sensing and quantifying error of the collector; $\Delta e3$ for recording error of the recorder; $\Delta e4$ for data deciphering error from the ground station (only valid for continuous parameters).

It can be seen that errors of FDRS are caused by various factors. Therefore, it is extremely difficult and impractical to apply traditional error compensation preprocessing methods to accurately estimate all error sources. It is especially more difficult for random interference error because of irreproducibility of live air flight.

Although errors of different types have different sources, they will be reflected in the final values. That is to say, they will present themselves in the form of deviations of actually measurement values—measurement noise. The fundamental purpose of flight data filtering preprocessing is to acquire more accurate measurement values. What is really of concern is the reflection of errors in the measurement values, not the concrete values of a certain error. To correct measurement values by estimating various errors is only an indirect method adopted to improve accuracy.

Based on the above-mentioned assumption, a unified error model of multisource flight data can be established as follows:

$$X'(k) = X(k) + V(k) \quad k = 1, 2, \ldots, n \quad (2.2)$$

In the formula, $X'(k)$ stands for the measurement value of parameter X at sampling time k; $X(k)$ for the true value of parameter X at sampling time k; $V(k)$ for the sum of error sources $\Delta e0$, $\Delta e1$, $\Delta e2$, $\Delta e3$, and $\Delta e4$ at sampling time k. That is to say, $V(k)$ is the general description of all error sources with unknown priori knowledge such as $E[V(k)]$ and $D[V(k)]$.

The mathematic description model for unified errors of flight data is a simple and practical method oriented to industrial application. It changes the traditional way to

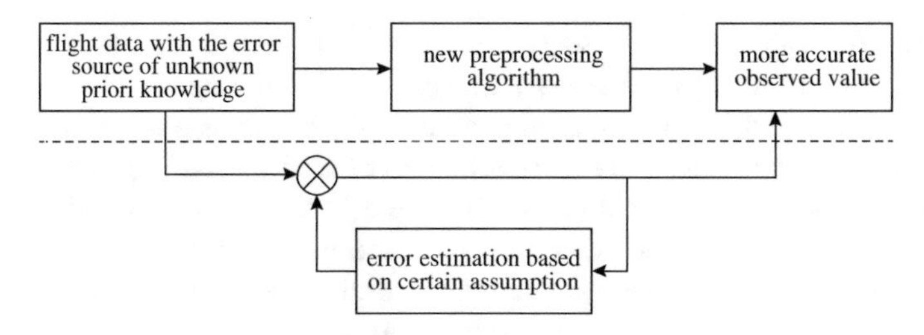

Fig. 2.1 Process contrast of two error compensation methods

analyze error compensation, eliminates many unnecessary assumptions, and avoids difficulty in estimating unknown priori errors.

Figure 2.1 shows the contrast of the different processing processes of the two error compensation methods. As shown in the figure, the block diagram below the dotted line reflects the preprocessing thinking of the traditional error compensation method. As in engineering practice, there is not such a thing as an ideal measured object, it is always the case that while using the traditional method, assumptive conditions cannot be met to guarantee the algorithm effective. The block diagram above the dotted line reflects the preprocessing thinking of the unified error compensation method. With unknown priori knowledge about the error source, the key to use the unified error compensation model to acquire more accurate measurement value lies in using filtering preprocessing algorithm independent of any priori knowledge.

2.1.2 Support Degree-Based Amnesic Fusion Filtering Algorithm

1. Problem Description

Current filtering algorithms for flight data preprocessing focus on a specific parameter, and thus have the following deficiencies:

(1) In the process of parameter filtering, judgment or weighting is fixed (as for an artificial neural network, the connecting weight is also fixed when training finishes). Thus its universality is poor;
(2) Once the parameter changes, it is complex, time-consuming, and inefficient to rebuild a model;
(3) Although the weight moving average filtering method takes into consideration the application of a larger filtering weight for information, lack of specific materialization foundation leads to insufficient utilization of such features as diversity and redundancy of FDRS-collected data, thus unable to validate the filtering data.

Compared with using one sensor, more reliable results can be acquired by utilizing multiple sensors for status parameters and using applicable fusing algorithm for them. And in this way, the measured object can also be more accurately identified. In recent years, parameter estimation based on multisensor data fusing technologies has aroused wide research interest. Diversity and redundancy of FDRS-collected and recorded data guarantee the information source for utilizing data fusing technologies. There are various recording media in modern FDRS such as primary recorder, secondary recorder, accident recorder, and QAR. Different recording media can provide multiple measured and recorded values for parameters with certain feature at the same time. Moreover, different information sources of the same recording media can also provide multiple measured and recorded values for parameters with certain feature at the same time, including pressure altitude and radio altitude, absolute altitude and vertical speed, heading provided by the attitude heading system and the inertial navigation system, etc.

At present, the common fusing algorithms can be classified into two types. The first type includes fusing methods based on priori knowledge(noise intensity, priori probability distribution, associated probability distribution, etc.), such as Bayes method based on statistics theory, optimal weight allocation methods based on planning, fusion methods based on Kalman Filters and its extended version, etc. As priori knowledge of measurement values is used for estimation, fusion works well. However, in engineering practices, these algorithms often fail because of inability to meet assumptive conditions. The second type includes fusion algorithms independent of priori knowledge, such as fusion methods based on relationship matrix, fusion methods based on relative distance and confidence distance, fusion methods based on closest measured distance, and consistency fusion based on support degree. These algorithms are oriented to engineering practices, applicable for occasions when priori knowledge of measurement values is unknown. However, with increasing measurement times, the increasing quantity of data greatly impacts the execution efficiency of these algorithms; meanwhile, the information provided by new data is submerged in the old data "ocean", and thus recursive algorithms lose corrective ability and cannot be updated when deviation from the true value occurs.

2. Measurement of Support Degree

For a sensor array consisting of n sensors, direct measurement is adopted to check static or gradually changing parameter X, i.e., $z_i(k) = X + v_i(k)$, $k = 1, 2, \ldots, n$. In this formula, $z_i(k)$ stands for the measurement value of sensor i at time k; X for the true value; $v_i(k)$ for measured noise at time k, with unknown priori knowledge of $E[V(k)]$ and $D[V(k)]$. Great difference between $z_i(k)$ and $z_j(k)$ means low mutual support degree of the measurement values of two sensors; otherwise means high. Exponential attenuation function is used to form a support degree matrix to express support degrees of these sensors at the same measurement time.

Definition 2.1 Support degree $\alpha_{ij}(k)$ of the measurement values of sensors i and j at time k is given by

$$a_{ij}(k) = \exp\left\{-a\left[z_i(k) - z_j(k)\right]^2\right\} \tag{2.3}$$

In the formula, a is an adjustable parameter. Using exponential attenuation function to express support degrees of measurement values of these sensors, the absolute values of 0 or 1 for support degrees in traditional methods are avoided. Therefore, the support degree of any two sensors at time k can be expressed by the support degree matrix SD(k):

$$\mathrm{SD}(k) = \begin{bmatrix} 1 & a_{12}(k) & \cdots & a_{1n}(k) \\ a_{21}(k) & 1 & \cdots & a_{2n}(k) \\ \vdots & \vdots & & \vdots \\ a_{n1}(k) & a_{n2}(k) & \cdots & 1 \end{bmatrix} \tag{2.4}$$

Obviously, for element i in support degree matrix $SD(k)$, if $\sum_{j=1}^{n} a_{ij}(k)$ has a relative large value, the measurement values of sensor i are more consistent with those of other sensors; otherwise, they deviate from those of other sensors.

3. Amnesic Factor

Definition 2.2 The function that determines the weight for new information and older information in the multistreaming time series, whose value varies with time, is called an amnesic function $A(t)$.

It is obvious that an amnesic function has monotonic property. That is, if $M \leq N, M, N \in R$, then $A(M) \leq A(N)$. This property is based on the fact that we would always believe that new information is more useful than older information, so the weight for the value of last measured data should be larger than that for the old one.

It should be noted that any monotonic nondecreasing function can be called amnesic function $A(t)$. However, the choice of $A(t)$ depends on the engineering issue itself. Here are some common amnesic functions.

(1) **Linear amnesic function**:

$$A(t) = at + b, a, b > 0$$

(2) **Exponential amnesic function**:

$$A(t) = e^{-at}, a > 1$$

(3) **Piecewise constant amnesic function**:

$$A(t) = \begin{cases} C_1, & 0 \leq t < M \\ C_2, & M \leq t \leq N \end{cases}$$

In the formula, $C_1 > 0, C_2 > 0, C_1 \leq C_2$ are constants and cannot be 1 at the same time; N represents current time, M is a parameter for amnesic depth. Obviously, M continuously increases with the ongoing measurements, but $N - M$ will remain constant. Piecewise constant amnesic function is simple in form and easy for engineering application. For the convenience of calculation, piecewise constant amnesic function is adopted for the applications in this section, with $C_1 = 0, C_2 = 1$.

4. Algorithm Descriptions

Definition 2.3 Consistency parameter of measurement values of sensor i with those of the other sensors at time k is given by

$$r_i(k) = \frac{\sum_{j=1}^{n} a_{ij}(k)}{n} \tag{2.5}$$

Obviously, $0 < r_i(k) \leq 1$.

Definition 2.1 reflects the approximation level of the measurement values of the two sensors at time k. Definition 2.3 reflects the approximation degree of the measurement values of sensor i with those of the other sensors (sensor i included) at time k.

Although $r_i(k)$ is large at one time, it does not mean that sensor i is reliable at any time of the whole observation range. The reliability of the sensor can be expressed by the consistency metrics at all measurement time. For example, $r_i(k)$ is very large at one time, and becomes very small at another, which indicates that consistency of sensor measurements is unstable. That is to say, the reliability of the sensor is poor during the whole measuring process. The intuition behind this is that the sensor with large mean value and steady variance of $r_i(k)$ time series should be assigned with bigger weight. Therefore, the two statistical concepts of mean value and variance value are used to study the reliability information contained in $r_i(k)$ series at different times.

Mean value of $r_i(k)$ series of sensor i at time k is

$$\overline{r_i(k)} = \frac{1}{k} \sum_{t=1}^{k} r_i(t) \tag{2.6}$$

Variance value of $r_i(k)$ series at time k is

$$\sigma_i^2(k) = \frac{1}{k} \sum_{t=1}^{k} \left[\overline{r_i(k)} - r_i(t)\right]^2 \tag{2.7}$$

Definition 2.4 $q_i(k)$ denotes the weight of the measurement value of sensor i at time k.

In the actual fusion process, measurement information of sensors with large mean value and small variance value (i.e., with high consistency and reliability) should be fully utilized. Therefore, weight $q_i(k)$ of the measurement value of sensor i should be in plus correlation with $\overline{r_i(k)}$ and minus correlation with $\sigma_i^2(k)$. To avoid $q_i(k)$ from being minus, linear function may be used to measure the final weight, i.e., $q_i(k) = [1 - \lambda\sigma_i^2(k)]\overline{r_i(k)}$, where λ is an adjustable parameter. By adjusting λ, the influence of $\sigma_i^2(k)$ on $q_i(k)$ will be changed accordingly.

So, the fusion estimation based on support degree can be expressed in this way:

$$\hat{\boldsymbol{X}}(k) = \frac{\sum_{i=1}^{n} q_i(k) z_i(k)}{\sum_{i=1}^{n} q_i(k)} \tag{2.8}$$

This formula is a consistent fusion algorithm. Amnesic function control item is further introduced and amnesic fusion algorithm based on support degree is denoted as the following formula:

$$\hat{\boldsymbol{X}}(k) = \frac{\sum_{i=1}^{n} w_i(k) z_i(k)}{\sum_{i=1}^{n} w_i(k)} \tag{2.9}$$

In this formula, $w_i(k) = [1 - \lambda\sigma_i^2(k)]\overline{r_i(k)}$ stands for amnesic fusion weight and can be calculated with the following formulas:

$$\overline{r_i(k)} = \begin{cases} \frac{\sum_{l=1}^{k} r_i(l)}{k} \, (k \leq m) \\ \frac{\sum_{l=k-m+1}^{k} r_i(l)}{m} \, (k > m) \end{cases} \tag{2.10}$$

$$\sigma_i^2(k) = \begin{cases} \frac{1}{k}\sum_{l=1}^{k} [\overline{r_i(k)} - r_i(l)]^2 (k \leq m) \\ \frac{1}{m}\sum_{l=k-m+1}^{k} [\overline{r_i(k)} - r_i(l)]^2 (k > m) \end{cases} \tag{2.11}$$

$$m = N - M$$

2.1.3 Application and Conclusion

Some flight data are chosen as samples. Sixteen groups of data about measurement records of "engine exhaust gas temperature" parameters are extracted from the FDMS primary recorder, QAR, and accident recorder. As shown in Table 2.1, the recorders are noted as "Sensor A, Sensor B, Sensor C," respectively, in accordance with the descriptions in data fusion field.

Table 2.1 Sensor measurement values

Measurement time	Measurement values (target value 900, unit: °C)		
	Sensor A	Sensor B	Sensor C
1	908.7	201.1	903.6
2	909.1	352.2	905.3
3	890.5	401.5	905.0
4	889.7	903.1	900.8
5	892.1	900.8	896.3
6	892.5	898.9	897.0
7	908.5	898.3	896.8
8	910.0	901.5	905.2
9	907.9	902.1	904.9
10	889.2	903.1	904.3
11	910.0	903.4	905.0
12	910.0	899.0	901.2
13	908.3	901.5	895.9
14	889.3	898.6	896.5
15	908.5	902.1	896.7
16	889.2	901.7	903.2

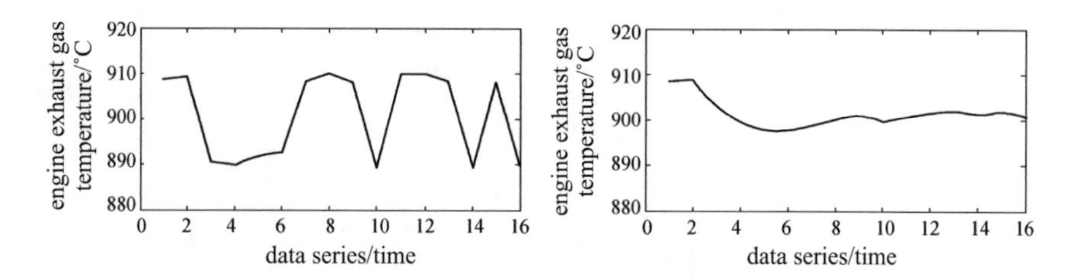

Fig. 2.2 Different effects of the original series and the filter series of Sensor A

Mean value fusion algorithm $X_N = \frac{1}{N}\sum_{i=1}^{N} x_i$ is used to preprocess the three sets of measurement data in Table 2.1, with n = 16. The different effects of the original series and the filter series of Sensor A and Sensor C are shown in Figs. 2.2 and 2.3, respectively.

As shown in Figs. 2.2 and 2.3, mean value fusion algorithm can be used to filter and smooth the data. The different effects of the original series and the filter series of Sensor B are shown in Fig. 2.4.

As shown in Fig. 2.4, the function of the arithmetic mean filtering method in preprocessing data is gradually degrading. There is an obvious deviation from the target value in the fourth filtering because of the following major reasons: during the iterative process, outliers are not eliminated in time and redundant information provided by multiple sensors is not efficiently utilized.

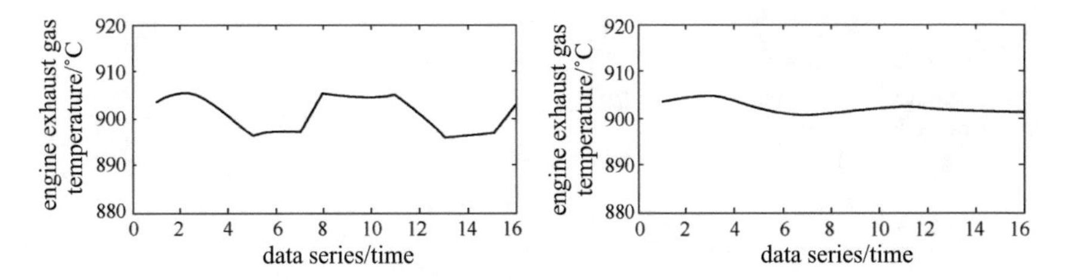

Fig. 2.3 Different effects of the original series and the filter series of Sensor C

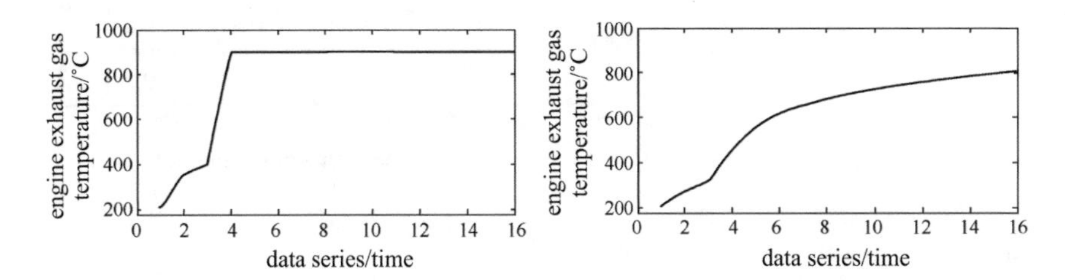

Fig. 2.4 Different effects of the original series and the filter series of Sensor B

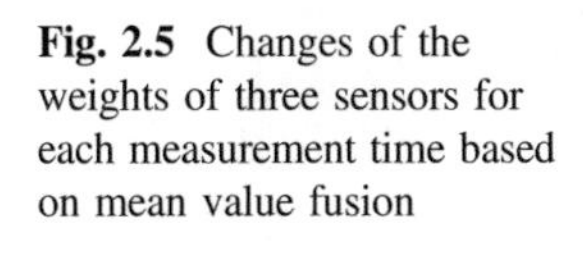
Fig. 2.5 Changes of the weights of three sensors for each measurement time based on mean value fusion

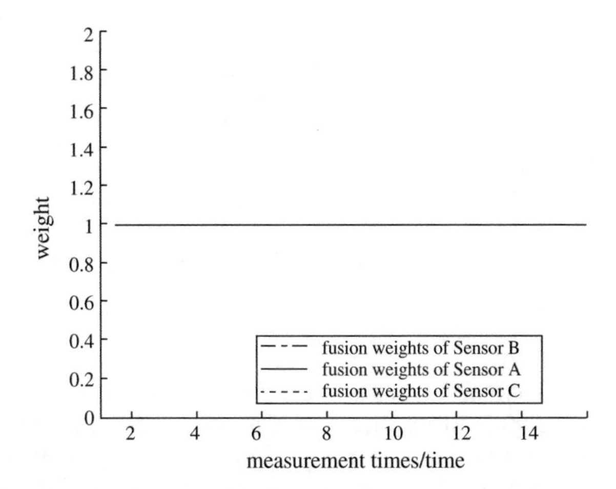

In the following section, mean value fusion algorithm, amnesic fusion algorithm based on support degree ("memory fusion" for short), and consistency fusion algorithm are used to filter and estimate the measurement values in Table 2.1.

For memory fusion and consistency fusion algorithms, the same weights used include measurement weight $\alpha = 0.85$, weight $\lambda = 0.35$, and memory weight $m = 5$. The changes of the weights of the three sensors for each measurement time are shown in Figs. 2.5, 2.6, and 2.7.

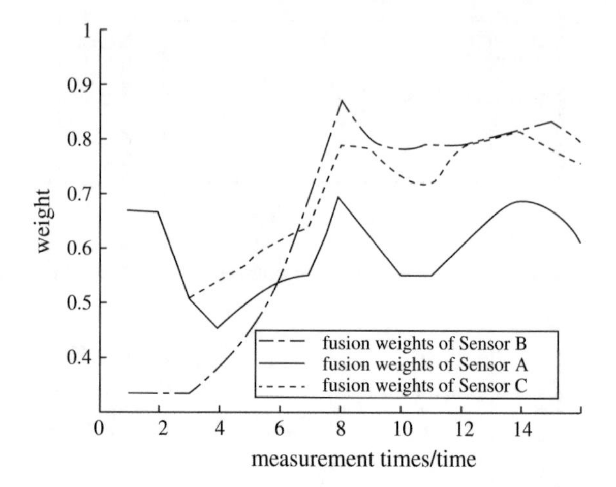

Fig. 2.6 Changes of the weights of three sensors for each measurement time based on memory fusion

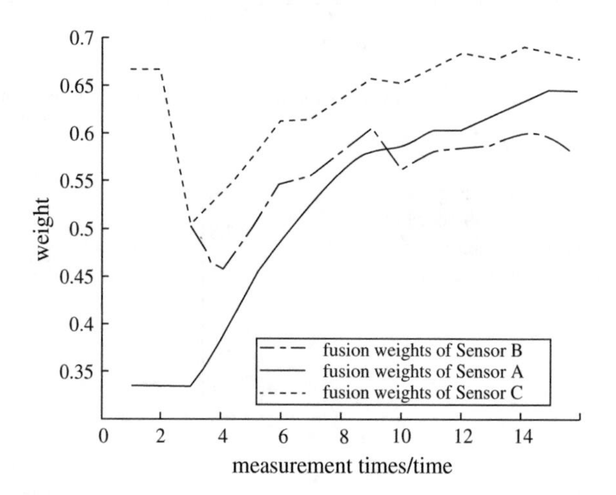

Fig. 2.7 Changes of the weights of three sensors for each measurement time based on consistency fusion

Obviously, for mean value fusion, a same weight (1 for all of the three sensors) is adopted to fuse measurement values; for consistency fusion, reliability information of the sensors provided by support degree matrix can be effectively used to allocate weights. But because of the influence of interferences, the weight of Sensor B cannot be effectively reallocated, still smaller than the weight of Sensor C in the end of measuring process; for limited memory fusion, based on consistency fusion, limited memory control item is also introduced. When the measurement value is updated for the eighth time, influence of interferences upon weight allocation is eliminated in time, and thus the weights of the three sensors are effectively and properly allocated again.

The fusion filtering effects generated by the three methods are shown in Figs. 2.8 and 2.9, where vertical coordinates refer to absolute error (°C), and horizontal coordinates refer to measurement times(time).

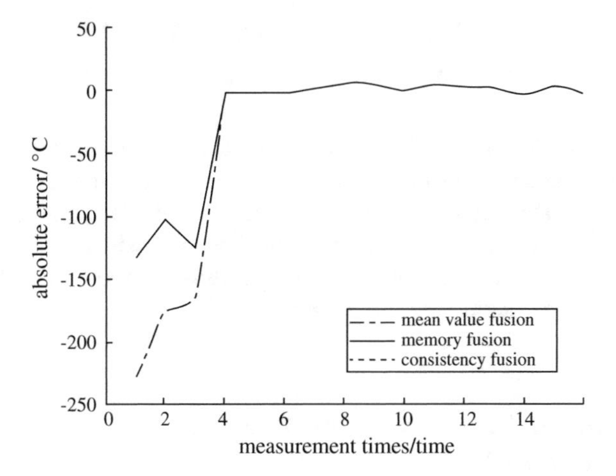

Fig. 2.8 Comparison of fusion filtering effects

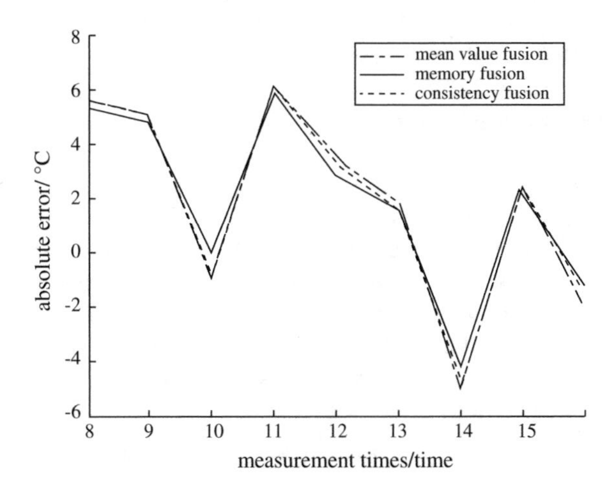

Fig. 2.9 Comparison of partially magnified fusion effects of measurement times 8–16

Figure 2.9 illustrates that as the measurement proceeds, amnesic control item plays an important role in adjusting the proper allocation of weights, thus further improving fusion accuracy. Judging from the changing process of absolute errors, with the application of amnesic fusion algorithm based on support degree, the filtering absolute errors are fewer than those in the other two methods as a whole, and they are steady and convergent.

2.2 Missing Flight Data Filling Method Based on Comprehensive Weighting Optimization

Because the operation of FDRS may be interfered by various factors, there may be missing values for recorded flight data, i.e., "missing records". Just like the processing of noise data, before further developing the flight data, effective data filling must be finished to ensure effective application of flight data. By combining comprehensive neural network and least square fitting, a method to fill the missing flight data under α condition is introduced.

2.2.1 *Modified Neural Network Model Based on Mixed Algorithm*

It is known that there are two problems in executing the algorithm of BP neural network: many partial minimal points and platform problem; and step size selecting problem. To solve these problems, the following methods are often adopted: mended BP (MBP), random optimization (RO), etc. Based on steepest descent optimization algorithm, MBP adopts directional searching method to choose the step size. Thus, low-quality partial minimal points may be converged. RO is not constrained by steepest descent, thus the searching direction can change randomly.

Modified neural network based on mixed algorithm adopts two-layer single-output network. In the network, measurement data is used as inputs, Sigmoid function and linear function are chosen, respectively, from the neuron at the hidden layer and the output layer, and corresponding fitting results are outputs. The learning algorithm of neural network combines the process of MBP and RO which are conducted by turns. When MBP result is judged to be a low-quality minimal point, RO is used; when RO result is judged to move close to a high-quality minimal point, MBP comes back.

1. Judgment of MBP Algorithm Running into a Partial Minimal Point

In the MBP algorithm, supposing $|\Delta R_{\text{emp}}(\xi(k))| = \delta(k)$, it gradually decreases with the increase of k. Supposing ε is a small plus, when k increases to a certain threshold, if $\left[\delta(k')/R_{\text{emp}}(\xi(k'))\right] < \varepsilon$, and $R_{\text{emp}}(\xi(k'))$ is still large, judgment of the algorithm running into a low-quality minimal point can be made.

2. Judgment of RO Moving from a Partial Minimal Point Close to a High-Quality Partial Minimal Point

In the RO algorithm, $\xi(k')$ is taken as initial value, and iterative computing is started from $k = 0$. Supposing $\delta(k)$ is the decreasing amount of experience risk for each iterative step, when k increases to a certain threshold, if the inequality $\sum_{k=0}^{k''} \delta(k) > \lambda R_{\text{emp}}(\xi(k'))$ exists, the judgment of moving is true. The left part of the inequality is the cumulative risk decreasing amount in the RO algorithm, and coefficient λ is a proper plus, $\lambda \in [0.1, 0.2]$.

2.2.2 Polynomial Fitting Model Based on Least Square Method

Supposing measurement data is $y(t)$, an n-order polynomial of a time variant is used during a certain period of time to describe:

$$y(t) = a_1 t^n + a_2 t^{n-1} + \cdots + a_n t + a_{n+1} \tag{2.12}$$

Supposing the measurement data for $N+1$ equal time intervals are $y_k, y_{k+1}, \ldots, y_{k+N}$ in turn, T refers to the interval, and $t_k = 0$ refers to the time of the initial measurement point, thus $t_{k+lT} = lT$. The following formula can be inferred:

$$y_{k+l} = a_1 (lT)^n + a_2 (lT)^{n-1} + \cdots + a_n (lT) + a_{n+1} \tag{2.13}$$

Suppose

$$a'_1 = a_1 T^n,\ a'_2 = a_2 T^{n-1},\ a'_3 = a_3 T^{n-2}, \ldots,\ a'_{n+1} = a_{n+1}$$

Then,

$$\begin{bmatrix} y_k \\ y_{k+1} \\ \cdot \\ \cdot \\ y_{k+N} \end{bmatrix} = \begin{bmatrix} 0 & 0 & \cdot & \cdot & 0 & 1 \\ 1 & 1 & \cdot & \cdot & 1 & 1 \\ \cdot & \cdot & & & \cdot & \cdot \\ \cdot & \cdot & & & \cdot & \cdot \\ N^n & N^{n-1} & \cdot & \cdot & N & 1 \end{bmatrix} \begin{bmatrix} a'_1 \\ a'_2 \\ \cdot \\ \cdot \\ a'_{n+1} \end{bmatrix} \tag{2.14}$$

Suppose

$$V = \begin{bmatrix} 0 & 0 & \cdot & \cdot & 0 & 1 \\ 1 & 1 & \cdot & \cdot & 1 & 1 \\ \cdot & \cdot & & & \cdot & \cdot \\ \cdot & \cdot & & & \cdot & \cdot \\ N^n & N^{n-1} & & & N & 1 \end{bmatrix}$$

$$Y = [y_k, y_{k+1}, \ldots, y_{k+N}]^T$$

$$X = \left[a'_1, \ldots, a'_{n+1}\right]$$

Substitute them into the formula (2.14) to get

$$Y = VX \tag{2.15}$$

Obviously, this is a contradictory equation group and it can be solved with least square method:

$$\hat{X} = \left[V^T V\right]^{-1}\left[V^T Y\right] \tag{2.16}$$

then,

$$\hat{Y} = V[V^T V]^{-1}[V^T Y]$$

Thus, it is easy to get the algorithm to calculate $\hat{y}_{k+l}(l = 0, \ldots, N)$, the estimated value of the n-order polynomial with $N + 1$ points at time l:

$$\hat{y}_{k+l} = [1 \quad l \quad l^2 \quad \cdot \quad \cdot \quad \cdot \quad l^n]\{[V^T V]^{-1}[V^T Y]\} \tag{2.17}$$

When $l > N$, the formula (2.17) is converted to use current measurement data to predict the estimated data at an unmeasured point. When $l = N + 1, N + 2, N + 3$, the values are status values for the follow-on first, second, and third time points. With the increasing distance between the predicted point and current measured data segment, prediction error will gradually increase. Thus, generally speaking, only the follow-on three points are predicted.

When $l = 0, 1, \cdots, N$, the formula (2.17) can be used to predict the estimated value for each measurement point. If the polynomial fitting algorithm is used for the whole data extent, the order of the polynomial would be too high, leading to instability of the solution of formula (2.16). Therefore, the polynomial moving fitting algorithm is adopted, where a small data segment is selected as a window with $(2M + 1)$ odd points.

Low-order polynomial fitting algorithm, a widely used central smoothing technique, is adopted within the window to estimate the status value of the central point of the window. With the moving of the estimating point, the window moves. For the initial and last M points, the window does not move, and the initial $(2M + 1)$ points and the last $(2M + 1)$ points are used, respectively, for polynomial fitting calculation.

If the order of the polynomial model is too low, the fitting would be rough; on the contrary, the data noise would be incorporated into the model. According to statistics, if χ^2 of the estimated parameters is close to its freedom level, the order is considered proper. When $y_k, y_{k+1}, \ldots, y_{k+N}$ of $N + 1$ measurement value is used to conduct n-order polynomial fitting, the estimated data is substituted into the following formula to calculate χ^2:

$$\chi^2 = \sum_{i=1}^{N} \left[\frac{y_i - (a_1 x_i^n + a_2 x_i^{n-1} + \cdots + a_n x_i + a_{n+1})}{\Delta y_i} \right]^2 \quad (2.18)$$

For χ^2 of a proper order, the degree of freedom $(N - n)$ should be taken as expected value. Therefore, start incrementally from $n = 2$ to calculate χ^2 of different orders. Judge the approximation level of $1 - p(\chi^2 < (N - n))$ to 0.5. In the actual calculation, F test is adopted to test whether the two matrix variances are equal. According to statistics, $F_n = \frac{(\chi_{n-1}^2 - \chi_n^2)}{\chi_n^2/(N-n)}$ complies with F distribution with freedom level as $(1, N - n)$, where n is current order, $N + 1$ is the number of fitting data, and χ_n^2 is χ^2 with n orders.

With the significant level α, determine the denial field of $F(1, N - n)$. If $F_n \le F_{\frac{\alpha}{2}}(1, N - n)$ or $F_n \ge F_{1-\frac{\alpha}{2}}(1, N - n)$, where $F_{\frac{\alpha}{2}}$ satisfies $P\{F \le F_{\frac{\alpha}{2}}\} = \frac{\alpha}{2}$, when F_n value is in the denial field, increase the order by 1, or current order is considered optimal.

2.2.3 Comprehensive Weighting Method to Fill the Missing Values

The fitting results from the above-mentioned two methods are weighted with the following formula:

$$X_i = \alpha X_{1i} + \beta X_{2i} \quad (2.19)$$

In the formula, X_i stands for the data after fitting at time i; α, β for weights; X_{1i}, X_{2i} for fitting values of BP neural network and least square method at time i, respectively.

Weights are chosen based on the following principles: when the data are too long and many data are missing, the fitting data weight of BP neural network is relatively large, i.e., $\alpha > \beta$; on the contrary, the fitting data weight of least square method is relatively large, i.e., $\alpha < \beta$. Weights can be chosen based on the results of massive data training.

This data filling method can take advantage of not only the excellent data fitting ability of the neural network, but also the least square method to make up for the deficiencies of the fitting results of a neural network with relatively few data, thus making the fitting data more reasonable and reliable.

2.2.4 Simulated Analysis and Conclusion

The missing values for pitch angle in the flight data are simulated, and the flight data curve including missing values is shown in Fig. 2.10. According to the figure, there are numerous missing values in the pitch angle data within 500 s, which has severe influence upon the data continuity and interferes with the follow-on data processing.

Figure 2.11 shows the effects of filling missing data through comprehensive weighting, neural network and least square polynomial fitting. It is shown that compared with the processing results of least square polynomial fitting and neural

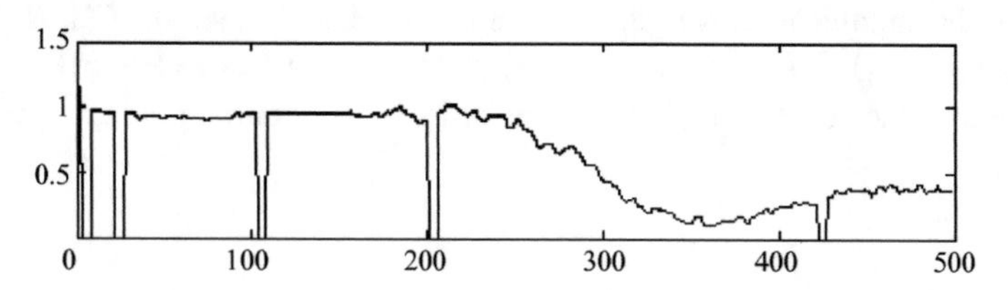

Fig. 2.10 Data curve generated by missing value simulation

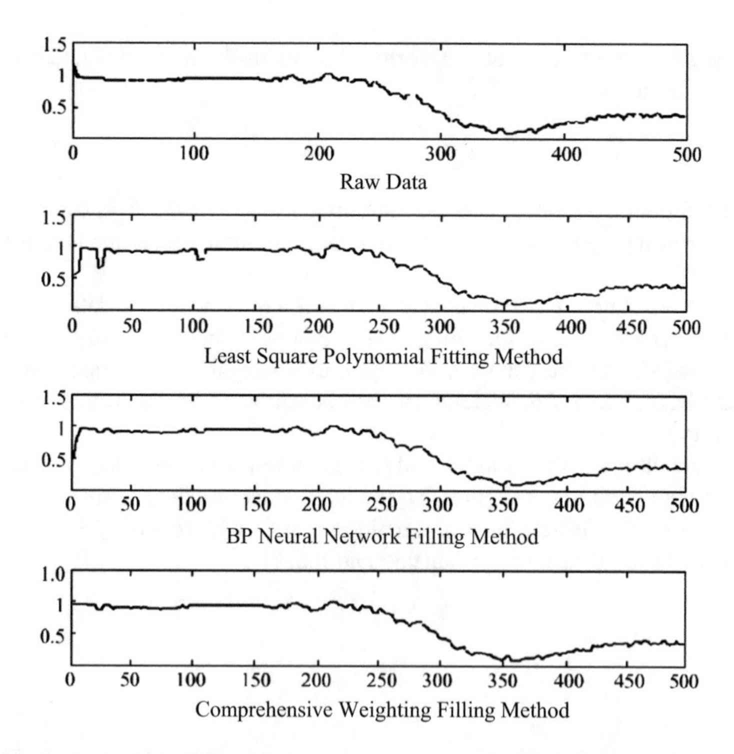

Fig. 2.11 Contrast of data filling effects

network, the data curve generated by comprehensive weighting data filling is more continuous and conformable to the raw and pure data. It reflects relatively true conditions of flight data and can effectively make up for the deficiencies of FDR data while guaranteeing data integrity, thus improving accuracy and reliability of the data.

2.3 Flight Data Extension Method Based on Virtual Sensor Technologies

For certain aircraft, the types and quantity of the parameters such as continuous and discrete data which can be collected by FDRS are relatively few and these parameters are far from enough to establish a relatively complete monitoring model. Moreover, when the analog sensor fails or drifts obviously, data cannot be acquired from this information source or even if data can be acquired, they cannot be used as valid monitoring data. Considering these problems, this section introduces the method to extend the flight data based on virtual sensor technology.

2.3.1 Overview of Virtual Sensor Technologies

A tolerance system is usually designed with redundant hardware, which means the need for additional hardware, such as computer, actuator, and sensor, so that the system can recover from partial damage. Additional sensors are often used to produce redundant output to conduct comparison in consistency. Although hardware redundancy is an important means for a tolerance system, it costs heavily in the aspects of funding, software, space for additional equipment, and maintenance. Virtual sensor-based software can partially replace some sensor hardware, thus reducing the requirements for cost and space. Moreover, it can reconstruct missing or hard-to-acquire measurement data with available data. As virtual sensors can be used to replace some hardware devices to reduce requirements for space and cost and meanwhile they have good performance, they have been widely applied in various fields such as communications, trouble diagnosis of sensor systems, robot systems, biochemical systems, target tracking, industrial manufacturing, medical care, noise control, and so on.

Virtual sensing, a signal processing technology, is used to estimate the response of a system's inability or difficulty to accommodate a physical sensor. Parameter estimation commonly starts from $Q = \{m_j : j = 1, 2, \ldots, n\}$, the set of measurement vectors m_j of $(m \times 1)$ to estimate the parameter vector f of $(t \times 1)$. The measurement vectors are the actual outputs of the physical sensor while n stands for the number of measurements. To implement the virtual sensor technology, it is assumed that $f = g(Q_s)$, mapping from Q_s, any subset of the measurement set Q, to the parameter vector f, is known. Thus, the virtual sensor technology divides the estimation problem into two separate parts. At first, calculate the parameter vector f (known as virtual sensor measurement) and the corresponding variance matrix according to the mapping $f = g(Q_s)$; then, calculate the parameter with smallest variance based on all the sensor measurements.

2.3.2 Virtual Flight Data Extension Based on Mathematical Model

The values of the virtual sensor come from the self-corrective real-time model of the engine and its theory is shown in Fig. 2.12. Suppose the engine model is a real-time component-level model with relatively high stability and dynamic precision, the status estimator conducts online estimation of the engine status by controlling input, based on the error between the output of the real engine and that of the engine model. Then the estimator takes corrected status value as benchmark for iterative calculation and thermal parameter calculation of the model.

While designing a virtual sensor, initialize the engine model first to make it consistent with the operational state of the engine. The input of the model includes the operational environment parameters (such as height, Mach number) of the engine and control volume (such as fuel), and meanwhile, the optimal estimated value of the status volume received by the model from the status estimator and the corrected initial values serve as benchmark to calculate the parameters of each section of the engine. In other words, the optimal estimated value serves as calculating benchmark for a component-level model. According to Huang Xianghua's research and calculation, after only one-step calculation, the component-level model can get thermal parameters close to those of a real engine, and thus the model can track a real engine.

Kalman algorithm is used for status estimation to linearize the component-level model first and then to establish a state space model of the engine status:

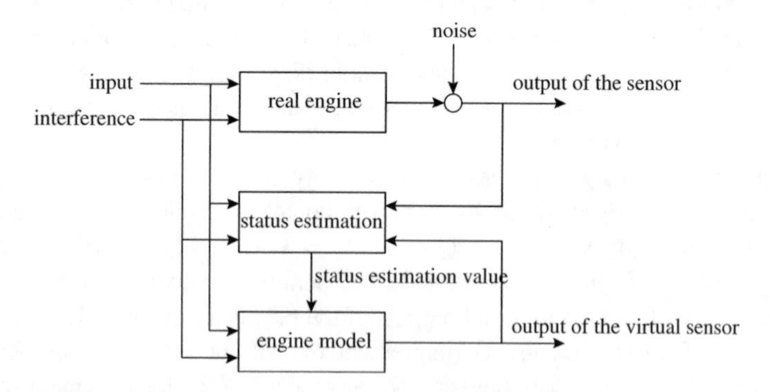

Fig. 2.12 Virtual flight data extension model based on the engine model

$$\delta_x = A\delta_x + B\delta_u + w \tag{2.20}$$

$$y = C\delta_x + D\delta_u + v \tag{2.21}$$

In the formula, w and v stand, respectively, for dynamic noise vector and measurement noise vector, which are irrelevant Gaussian white noise with the mean value as zero. Similar normalized parameters are adopted for status value, input value, and output value, which means a linear model with small deviance is established with the stable operating point of the engine as benchmark.

The discrete optimal filtering estimation algorithm is as follows:

$$\hat{\delta}_x(k+1) = \hat{\delta}_x(k+1,k) + K(k)(\delta_y(k) - \hat{\delta}_y(k)) \tag{2.22}$$

Under real-time control, it takes a long time to calculate the optimal filtering estimation value, even longer than the permitted sampling time. While a constant gain suboptimal filter is used for filtering, i.e., $K(k) = K_d$ (K_d stands for the stable constant gain matrix for optimal filtering), this filtering is suboptimal before it gets stable state. As Kalman filter itself is robust in nature, the constant gain suboptimal filter can relatively effectively estimate the status value.

Mathematic model is an important tool for modern aeroengine manufacturing, testing, and operation. Precise mathematic model guarantees the precise extension of the virtual flight data based on the engine model. However, because of technological security, it is rather difficult to thoroughly master the precise mathematic model that can fully reflect the operational state and process of an engine, giving rise to the difficulty in fault analysis in the process of teaching, research, and application. At present, the appearance of identification methods such as least square of an engine model based on flight data and Volterra series model provides a foundation for the research of precision engine modeling. However, these methods are not practical enough and must be aided greatly by subject experts. This is one of the important topics that need to be studied.

2.3.3 Virtual Flight Data Extension Based on BP Network

1. Problem Description

Usually, there is certain mapping relationship among various parameters which describe the aircraft condition. So long as there is a precise system mathematical model, the required virtual flight data can be estimated with known data of one or more parameters. However, as the mathematic models of each subsystem aboard the aircraft are usually complex nonlinear models, it is relatively complicated to use traditional modeling method to establish a model and the accuracy of estimated data will be influenced.

BP neural network, a parallel computation model, has many unique advantages: excellent nonlinear mapping ability; few requirements for empirical knowledge about the modeling object; no need for knowledge about the structure, parameters and dynamic features of the modeling object; and only with input and output data of the object, the relationship between the input and output of the network can be acquired through the learning function of the network itself. Given the superb advantage of the nonlinear feature of the neural network, the method to estimate virtual flight data based on BP neural network will be introduced in the following paragraphs.

2. Algorithm to Acquire Virtual Data Based on BP Network

The basic tenet to acquire virtual data based on neural network is: with data of all parameters except those to be estimated as input of the network, and data of the parameters to be estimated as output, utilize the known data training network in the system and when the network meets requirements, input into the network the parameters to be estimated and other parameters to acquire estimated values of the virtual data.

1) Data Normalization

BP neural network adjusts the connecting weight according to gradient descent method to minimize the error function. During the process of adjusting the connecting weight, when the input x_i is relatively large, the input of next-layer node transferring function is relatively large. The output becomes saturated, close to 1. To meet the requirements of BP neural network, the input data must be normalized. Fuzzy membership normalization method is adopted. The following formula is often used for data normalization:

$$\mu_i(x) = \frac{k(x_{\mathrm{i}} - \alpha_i)^2}{1 + k(x_i - \alpha_i)^2} \tag{2.23}$$

In the formula, x_i stands for input sample value, α_i for the maximum of the input sample value, k for proportional coefficient.

While the formula is used to normalize the flight data, only the degree of difference between the recorded data and the estimated data is taken into consideration, and the deviation direction is not considered. Therefore, some information will be lost after normalization processing. To reduce lose of information, the formula is modified as follows:

$$\mu_i(x) = \frac{k(x_i - \alpha_i)|x_i - \alpha_i|}{1 + k(x_i - \alpha_i)^2} \tag{2.24}$$

In the formula, k is a parameter determined by the environment. The flight data can be normalized into the segment [−1,1] by selecting proper value of k.

2) Virtual Data Acquisition

A three-layer BP neural network is adopted. Suppose *m, n,* and *u* stand for the node number of the input layer, output layer, and hidden layer, respectively. The node output function of the output layer is

$$c_j = f\left(\sum_{r=1}^{n} V_{rj} \cdot b_r + \theta_j\right) \; (j = 1, \ldots, n) \tag{2.25}$$

In the formula, V_{rj} stands for the connecting weight between the hidden layer and the output layer, b_r for the node value of the hidden layer, θ_j for the node threshold of the output layer, and c_j for the node value of the output layer.

The node output function of the hidden layer is

$$b_r = f\left(m \sum_{i=1}^{m} W_{ir} \cdot a_i + T_r\right) \quad (r = 1, \ldots, u) \tag{2.26}$$

In the formula, W_{ir} stands for the connecting weight between the input layer and the hidden layer, α_i for the node value of the input layer, and T_r for the node threshold of the hidden layer. The action function $f(x)$ usually takes the form of the sigmoid function $f(x) = (1 + e^{-x})^{-1}$.

The learning process of the neural network is: While the neural network is learning, if the error between the value of the output layer and the expected value is larger than the permitted error, the network will adjust the connecting weight and the threshold of the nodes in the connecting layer. In this way, the error of the node in the output layer will be reversely propagation for the input layer to distribute the error to each connecting node, thus figuring out the error reference of each connecting node. Then adjust them correspondingly according to the weight and threshold of each connecting node to enable the network to meet the output requirements, thus realizing the mapping of $A^{(k)} \to C^{(k)}(k = 1, 2, \ldots m)$, where $A^{(k)} = (a_1^{(k)}, a_2^{(k)}, \ldots, a_m^{(k)})$, $C^{(k)} = (c_1^{(k)}, c_2^{(k)}, \ldots, c_n^{(k)})$, $a_i^{(k)} \in R, c_j^{(k)} \in R$, ($R$ is real number field.)

The concrete calculation includes the following steps:

Step 1: set randomly a small value between 0 and 1 for W_{ir}, T_r, V_{rj}, and θ_j.

Step 2: input the value of $\alpha_i^{(k)}$ into the node of the input layer and calculate forward b_r and c_j in turn.

Step 3: calculate the error between the node output value and the expected output value of the input layer $d_j : d_j = c_j \cdot (1 - c_j) \cdot (c_j^{(k)} - c_j)$.

Step 4: back error propagation for the nodes of the hidden layer e_r: $e_r = b_r \cdot (1 - b_r) \cdot \left(\sum_{j=1}^{n} V_{rj} \cdot d_j\right)$.

Step 5: to reduce vibration during the learning process, Rumelbart with inertial impulse technology is adopted for adjusting the weight and threshold.

Inertial impulse is added to filter the high-frequency vibration during the learning process, thus getting the maximal learning rate to accelerate learning. Here, the general form for adjusting the weight is

$$\Delta V_{rj}(t+1) = -\lambda \frac{\partial E}{\partial V_{rj}} + \eta \Delta V_{rj}(t) = \sum_{k=1}^{m} (\lambda b_r d_j) + \eta \Delta V_{rj}(t) \quad (2.27)$$

$$\Delta W_{ir}(t+1) = -\beta \frac{\partial E}{\partial V_{rj}} + \delta \Delta W_{ir}(t) = \sum_{k=1}^{m} (\beta a_i e_r) + \delta \Delta W_{ir}(t) \quad (2.28)$$

The general form for adjusting the threshold is

$$\Delta \theta_j(t+1) = \sum_{i=1}^{m} \lambda d_j + \eta \Delta \theta_j(t) \quad (2.29)$$

$$\Delta T_r(t+1) = \sum_{k=1}^{m} \beta e_j + \delta \Delta T_r(t) \quad (2.30)$$

In the formula, λ and β stand for learning rates, usually between 0 and 1; η and δ for dynamic factor; E for the sum of squared error in the whole training set, i.e.,

$$E = \frac{1}{2} \sum_{k=1}^{m} \sum_{j=1}^{n} (C_j^k - C_j)^2, \quad E = \sum_{k=1}^{m} E_k \quad (2.31)$$

Step 6: repeat Steps 2–5 and terminate training when the error d_j meets the requirements or becomes 0.
Select the flight data from the same sortie as training samples. With the data of known parameters related to missing parameters as input and the data of missing parameters to be estimated as output, conduct the above-mentioned iterative process repeatedly until the requirements are met. In this way, a matching neural network model can be acquired.

Step 7: input the data of relevant parameters at the same moment of the to-be-estimated data into the trained network. The output value is the estimated value.

3. Experiment and Result Analysis

Take an engine system for example, adopt the given three-layer BP neural network model to simulate, with 6500 s data recorded by the FDRS as data samples, where the first 6000 s data as training samples, and the remaining 500 s data as test samples; the throttle angle and rotating speed of the engine as input value of

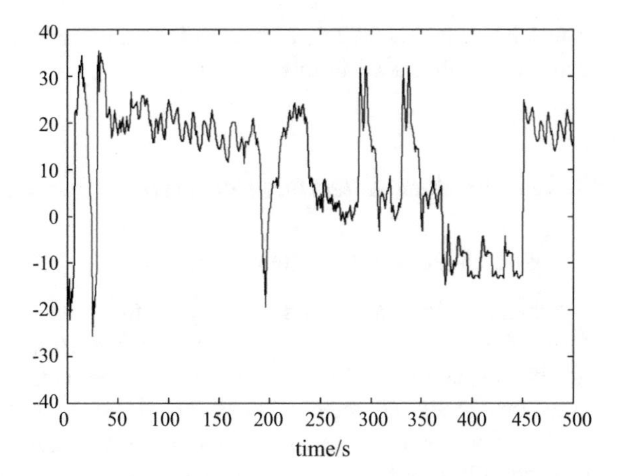

Fig. 2.13 Curve of the error between the estimated value and the actual value of the virtual data

the model, exhaust gas temperature as output value; 2 input nodes, 25 hidden nodes, and 1 output node.

Standardize the data sample with $k = 0.00001$, $\alpha_1 = 110$, $\alpha_2 = 8000$, and $\alpha_3 = 900$. Then, input the first 6000 s standardized data into the above-mentioned network model for training, with 0.0001 for E, the sum of squared error, learning rate $\lambda = \beta = 0.01$, dynamic factor $\eta = \delta = 0.75$. After 682 iterations, the network meets the requirements. Input the test samples into the trained network to estimate the virtual parameters. Figure 2.13 shows the curve of the error between the estimated value and the actual value of the 500 s virtual data with maximum relative error as 0.038%.

The experiment shows that when the sensor for exhaust gas temperature has trouble, resulting in invalid data, it is feasible to acquire the virtual data based on BP network to meet application requirements.

2.4 Self-expanding Genetic Algorithm for Feature Selection in Monitoring Flight Data Capacity

It is an issue of classifying or clustering model identifications in nature to utilize flight data to monitor the aircraft equipment condition or diagnose a trouble. The key of the issue lies in the establishment of mapping from the equipment data to trouble symptom space. The condition can be better studied by selecting the features at different conditions represented by the flight data. Therefore, the key feature parameters distinguishing different patterns are critical input. Generally speaking, as flight data are multivariable high-dimensional time series, numerous feature parameters are associated with each other. Too many input values used for model identification will reduce operability and effectiveness of the results. Therefore, an important preprocessing task is to optimize representative feature parameter sets.

The self-expanding genetic algorithm for feature selection in monitoring flight data capacity is introduced in this section.

2.4.1 Feature Selection and Genetic Algorithm

1. Research Domain of Feature Selection

Objective data can be indicated by vectors, and these vector sets may take the form of certain linear or nonlinear manifold structure. Because of the strong association caused by the model and other potential factors, there are always greatly redundant observing vectors. To explore linear or nonlinear structure from limited discrete dot sets (possibly with certain noise interference) to classify, clustering and data visualization becomes a challenging problem in supervised learning, that is, feature dimension reduction problem which needs to be solved for statistical pattern identification. Dimension reduction methods are divided into two categories: feature selection and feature extraction. Feature selection is subdivided into feature filtering and feature checking while feature extraction includes linear and nonlinear feature extraction. The optimal methods for feature filtering include exhaustive method, branch and bound algorithm, feature selection based on genetic algorithm, etc. Linear feature extraction methods include principal component analysis, Fisher identification analysis, factor analysis, multidimensional scaling, etc. Nonlinear feature extraction methods consist of principal component analysis based on nucleus and identification analysis based on nucleus.

Optimal subset needs to be established for feature selection by starting from a null set to establish d features incrementally, or starting from measured full-value set to delete redundant features gradually. The methods can be divided into two categories: filter and wrapper, i.e., filter and checking, as shown in Fig. 2.14.

Based on the optimal feature subset acquired by identification function, feature filtering only depends on the statistical feature of training samples, independent of the learning algorithm of the classification device. However, feature checking estimates the strong and weak points of each feature subset for actual classification effects based on the learning algorithm of the classification device. It is not only

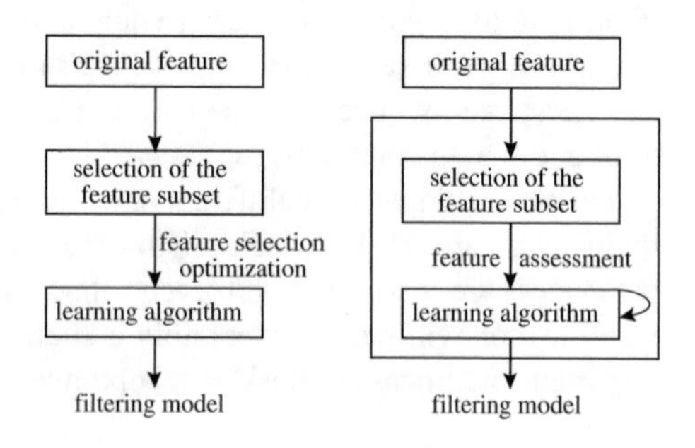

Fig. 2.14 Two feature selection models

related with the statistical feature of training samples, but also closely related with the statistical feature of test sample and learning algorithm.

Feature selection can be described mathematically: given measurement values for p variants, find the optimal subset of d. Optimization is conducted based on the set of subset X_d of d, that is to find subset $\overline{X}_d$.

$$J(\overline{X}_d) = \max_{X \in X_d} J(X) \tag{2.32}$$

J rule function is usually based on measurement of distance or distribution difference. Formula 2.33 is the possible number of subsets. To select 10 features from 25 variants means that 3,268,760 subsets should be considered and optimal rule should be calculated for each subset.

$$n_d = \frac{p!}{(p-d)!d!} \tag{2.33}$$

While monitoring the aircraft condition, there are some special points for reducing the dimensions of feature parameters of flight data. First, with limited trouble samples from flight data, to design classifier with numerous features does not conform to actual requirements. Second, sometimes, intuitional physical meaning cannot be acquired through feature selection methods. Therefore, it is difficult to figure out the influence of each feature upon the classification. Third, as the number of the recorded parameters in the flight data is not the minimal dimension by which pattern cannot be distinguished, eliminating redundant or interference features helps to identify the aircraft condition model accurately. Given the above-mentioned factors, feature selection becomes the primary means to reduce the dimensions of feature parameters of flight data.

2. Feature Selection Based on Genetic Algorithm

Genetic algorithm (GA) is a highly parallel, random, and self-adaptive search algorithm based on natural selection and natural heredity, which stimulates the principle of survival of the fittest during the biologic evolution process to conduct simultaneous optimization of multiple parameters and populations. During the research process of evolution calculation, Holland proposed bit string coding technology which not only applies to mutation operation but also to crossover operation, emphasizing that crossover should be used as the primary genetic operation. After applied to the research of self-adaptive performance of natural and artificial systems, the algorithm, known as genetic algorithm, has been successfully applied to such aspects as optimization and learning. The algorithm includes genetic algorithm, evolution strategy, evolution planning, and genetic program design.

Genetic algorithm is widely used in feature selection. However, with new samples, a standard algorithm is usually adopted in feature selection for all samples. With relatively great interference of new samples, the original speed and results of feature selection will be influenced. Therefore, self-expanding genetic algorithm based on increasing of samples is put forward. When there are new samples, excellent partial features are reselected based on the original feature selection to avoid repeated feature selection for all samples, thus the fitness level is increased and iteration times are reduced greatly.

2.4.2 *Self-expanding Genetic Algorithm*

Selection of features indicating aircraft troubles based on flight data has the following characteristics: first, current flight data may not include all types of trouble states; second, new trouble modes included in the trouble data samples will influence the results of original feature selection. With accession of new data in the flight database, there must be many new positive and negative samples. Repeated optimization will lead to complicated calculation and the evolution features of the genetic algorithm will not be fully utilized.

1. Structure of Self-Expanding Genetic Algorithm

Self-expanding genetic algorithm can be used to solve the above-mentioned problems effectively. The basic principle of the algorithm is to use acquired features to classify new samples. If classification accuracy decreases, select a certain proportion of excellent individuals according to fitness level of the individuals from the population acquired after the previous execution of the genetic algorithm. These selected individuals will be used as the initial individuals for this execution while other individuals in the initial population will be generated randomly as shown in Fig. 2.15. The self-expanding genetic algorithm mainly includes the following steps:

(1) Initialize the control parameters such as population number, threshold, mutation probability.
(2) Generate the initial population G randomly, calculate the fitness level f of each individual in the population and the average fitness level of the population.
(3) Use selection, crossover, and mutation operators to generate filial generation and then evaluate the filial generation, i.e., calculate the fitness level of the filial generation. If the fitness of the newly acquired filial generation is better than the parent's, use the filial generation to substitute the parent, otherwise accept the filial generation.
(4) Judge the converging condition—If the value difference of the average fitness level between neighboring generations is smaller than given threshold, searching is terminated. Sort the results according to fitness values and select the result with the highest fitness value as the optimal solution.
(5) Incorporate the newly acquired samples into the original sample set and classify them with acquired features. If classification accuracy does not decrease, current features are valid; whereas, select a certain proportion of excellent individuals according to the fitness level of the individuals from the population

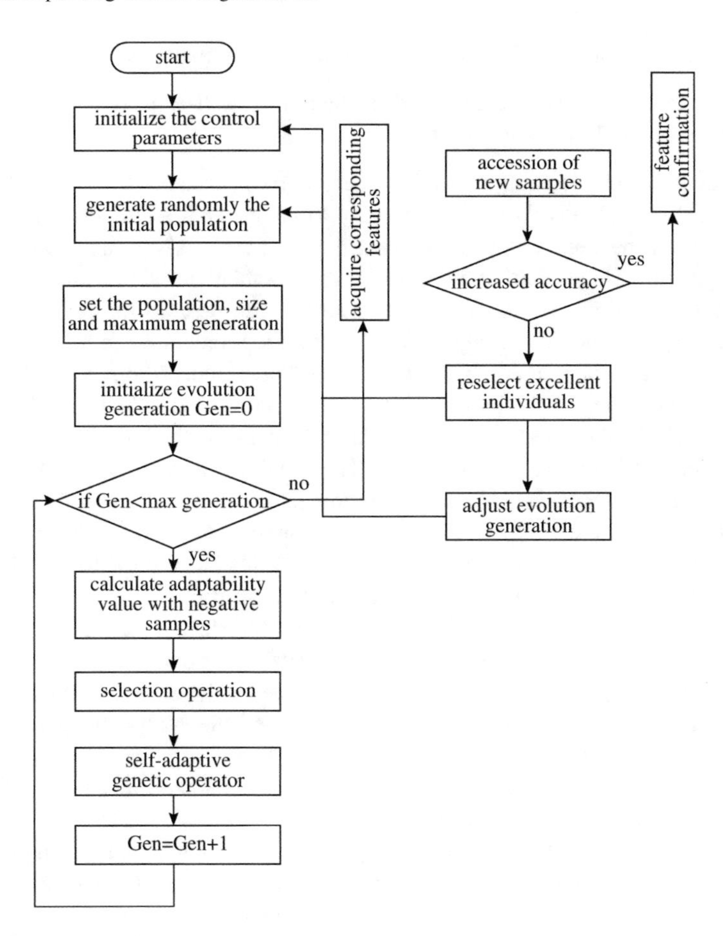

Fig. 2.15 Process of self-expanding genetic algorithm

acquired after the previous execution of the genetic algorithm. These selected individuals will be used as the initial individuals for this execution while other individuals in the initial population will be generated randomly.

(6) Adjust the maximum generation $n_1 = n_0 * (s_1/s_0)$, where n_0 stands for the original maximum generation, s_1 for current sample number, and s_0 for original sample number. Do not terminate the procedure until one of the following conditions is satisfied: first, the evolution generation reaches the maximum given evolution generation; second, the maximum fitness level is no less than the fitness level of the original rule and the maximum fitness level will not change during the given N-generation evolution process.

2. Coding, Evaluating, and Operator Designing of Self-Expanding Genetic Algorithm

The operation, representation, and evaluation of self-expanding genetic algorithm include the following aspects:

(1) Coding. Binary coding, a coding method most widely used in the genetic algorithm, is relatively simple and easy to understand. Although binary coded string will be very long while solving high-dimensional optimization problem, the length of flight data trouble sample itself is relatively limited and cannot be expanded infinitely. Therefore, binary strings can be used for coding.
(2) Generation of the initial population. The initial population can be generated in this way: generate randomly a binary string (individual) with the length D, with d as the number of "1", and continuously produce M individuals of this sample. M stands for the population size, which influences the final result and execution efficiency of the genetic algorithm. Too small M may easily lead to partial optimization, thus influencing its optimization performance; otherwise, the computation will be very complicated. Suppose $M = tn$, where t is a real number between 1 and 2, then at least n samples are needed to distribute the initial population in the problem space rather evenly.
(3) Confirmation of evaluation functions. During the early phase of the algorithm operation, the fitness levels differ greatly, possibly resulting in prematuration in the selection process. But during the late phase of the algorithm operation, the individual fitness levels may be very close, and the competitiveness of most individuals has no bearing on the optimal individual, thus possibly entering into random selection process. Linear scaling can be adopted to adjust individual fitness level in different operation phases.
(4) Chromosome selection operator. Proportional selection is a random playback sampling method in which the probability for selecting each individual is directly proportional to its fitness level. Because of random operation, this method has relatively large selection error, and sometimes even the individual with high fitness level cannot be selected. Therefore, optimal saving tactics is adopted to ensure that the optimal individuals acquired so far cannot be destroyed by genetic calculation such as crossover and mutation. This tactics is important to guarantee the convergence of the genetic algorithm. When the fitness level of the worst individual in current population is lower than the overall fitness level of the optimal individuals, the best individual is used to replace the worst individual in current population.
(5) Designing of chromosome crossover operator. The global search function of the genetic algorithm is directly dependent upon crossover algorithm, and LOX crossover operator should be adopted. The operator can be constructed in the following way: first, select a pair of parent individuals randomly; then, select a feature randomly, with which to conduct m operations for each parent individual to get a new string. Move leftward or rightward for L positions (L is determined randomly), but it is necessary to keep stable the distance between the operations corresponding to selected work. Use the operations corresponding to the features not selected in the second parent individual to replace

the original individuals one by one according to their original relative sequence to generate a descendant individual.

(6) Crossover probability and mutation probability. Self-adaptive crossover and mutation probability selection method is adopted. $f_{\max}$ stands for the maximum fitness level of the population, $\bar{f}$ for the average fitness level of the population, f_c for the higher fitness level in the two strings participating in the mutation, and f_{m} for the fitness level of the mutating string. The self-adaptive adjustment of p_c and p_m is inversely proportional to the convergence of the algorithm, and can effectively prevent the convergence from being partially minimal. The higher the fitness level of the individual is, the smaller the corresponding p_c and p_m are, and thus good evolution results are kept. The corresponding p_c and p_m can be acquired with the following formula in which $k_i = (i = 1, 2, 3, 4)$ stands for adjustable parameter which can be set with reference to typical genetic algorithm.

$$p_c = \begin{cases} k_1(f_{\max} - f_c)/(f_{\max} - \bar{f}_c), \; f_c \bar{\geq} f \\ k_2, f_c < \bar{f} \end{cases}$$
$$p_m = \begin{cases} k_3(f_{\max} - f_m)/(f_{\max} - \bar{f}_m), \; f_m \bar{\geq} f \\ k_4, f_m < \bar{f} \end{cases} \tag{2.34}$$

2.4.3 *Case Verification and Assessment*

With the engine of a type of aircraft as the research object, continuous flight data include speed of the high-pressure rotor speed, throttle position, vibration speed of the aft pod, exhaust gas temperature; discrete type flight data include metal element concentration of lubricating oil. With the typical trouble of abnormal vibration value of the engine for example, verify the accuracy of self-expanding genetic algorithm. Data samples are taken from abnormal engine vibration values. Severe bearing attrition results in engine replacement ahead of schedule. Figure 2.16 shows a group of flight data samples under the stable airborne condition. The data samples are speed of the high-pressure rotor of the engine, throttle position, vibration speed of the aft pod, and exhaust gas temperature in sequence. The unit for horizontal coordinates is second (s).

Select 26 normal positive samples and 76 negative samples of the engine with the length of chromosome lchrom = 12, the size of the population popsize = 150, maximum generation Max(Gen) = 300. Figure 2.17 shows the curve of the evolution process with severe engine bearing attrition, where horizontal coordinates stand for evolution generation and vertical coordinates for the fitness level. The fitness level of the optimal individual is acquired: fitness = 68.3653258 with presentation string of the chromosome as "011011000100". According to the

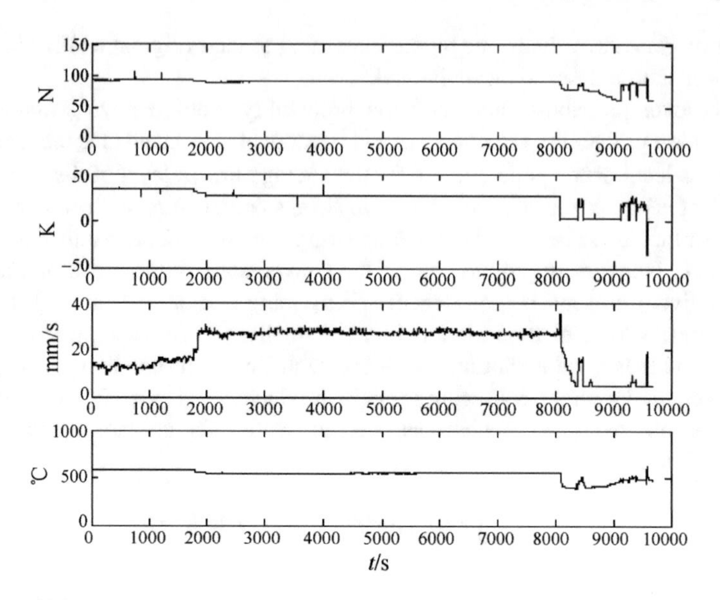

Fig. 2.16 Flight data samples

Fig. 2.17 Actual evolution results

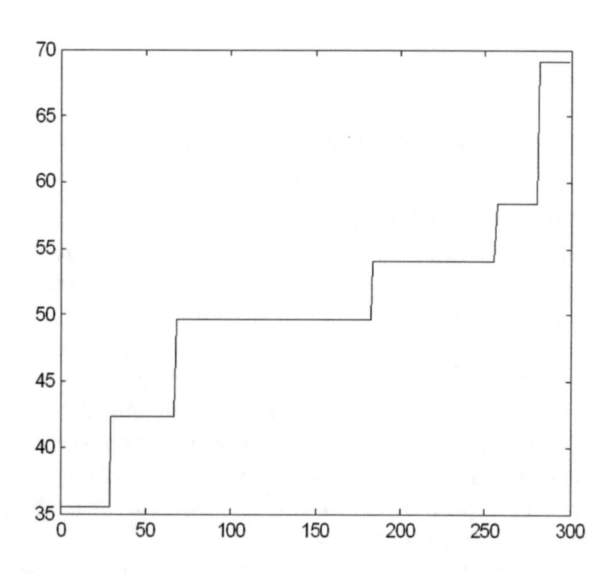

classification generation rule, the above rule can be explained in this way: ($C_3 < 521.97 \&\& C_5 < 58.8115 \&\& C_8 > 44.8745$). Based on corresponding expertise, the physical meaning can be interpreted in this way: Under the stable airborne condition with rotor speed larger than 97.153% or smaller than 99.234%, percentage of vibration speed change per second larger than 24.454%, percentage of exhaust gas temperature change per second larger than 12.235% and smaller than

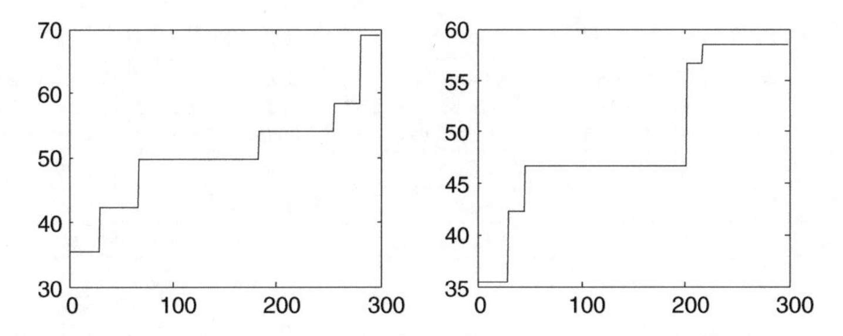

Fig. 2.18 Contrast between standard genetic algorithm and self-expanding algorithm

34.564%, the features acquired are rotor speed, vibration speed, and exhaust gas temperature, corresponding to the severe engine bearing attrition phenomena. The feature selection results completely conform to the knowledge of frontline experts.

Add 10 positive examples and 5 negative examples, thus increasing the number of positive and negative examples to 36 and 81, respectively, with other parameters unchanged. Figure 2.18 shows the contrast between self-expanding algorithm and standard genetic algorithm. It is obvious that self-expanding algorithm conducts efficient feature selection based on the original feature selection, and thus the starting point of the fitness level in the evolution process becomes higher, the iteration times greatly decrease and interference of noise data is eliminated.

Use a binary-class support vector machine (SVM) to conduct classification with the three acquired features. The samples are taken from the flight data of a type of aircraft accumulated in recent 2 years and each sortie corresponds to a group of samples. Table 2.2 shows the classification results, where SGA stands for self-expanding genetic algorithm, GA for standard genetic algorithm, and SVM for binary-class support vector machine. The table shows that because of the influence of interference parameters, when the samples increase to certain number, the classification accuracy decreases with the adoption of only support vector machine (SVM). The classification accuracy of feature sets after feature selection of genetic algorithm increases, which indicates that feature selection greatly reduces the influence of interference factors and thus classification accuracy is enhanced. While using the given self-expanding genetic algorithm for feature selection, the number of features further decreases, but the classification accuracy is greatly enhanced.

2.5 Chaotic Property Analysis of Flight Data

Phase space reconstruction usually uses a sequence to analyze the dynamic property of the original system and after selecting effective features, analyzes the property of the system from which the data comes, including chaotic prediction, estimation of dynamic invariants, characterizes discrimination of chaotic signals, etc. Actually, the time series generated by the flight data often reflects the dynamic property of the

Table 2.2 Contrast of classification accuracy with or without adoption of feature selection

Method	Feature	Classification accuracy with 800 groups of samples	Classification accuracy with 1500 groups of samples	Classification accuracy with 2000 groups of samples	Classification accuracy with 3000 groups of samples
SVM	12	78.52	81.38	83.35	85.57
GA+SVM	8	83.82	85.25	87.49	91.12
SGA+SVM	7	93.17	94.26	95.37	96.54

original system. Phase space reconstruction method can be used to analyze the features of the flight data qualitatively and quantitatively. In this section, based on phase space reconstruction of the original flight data, typical flight data are analyzed with such methods as power spectrum graphs and Maximum Lyapunov Exponent. It is proven that they have typical chaotic property, thus exploring a new way to further study flight data processing technologies.

2.5.1 Mathematical Description of Phase Space Reconstruction of the Chaotic Series

1. Phase Space of the Chaotic Series

Chaos, a kind of aperiodic, macroscopic, and spatiotemporal behavior, is generated by a nonlinear dynamic process inherent in a system. It integrates skillfully the apparent disorder and inherent regularity and reflects the inherent randomness of a nonlinear dynamic system. It is a seeming irregularity and random-like process existing in a deterministic system. A system with both sensitivity to the original values and aperiodic movement is a chaotic system.

The state of a system at certain moment is known as phase, and the geometric space determining the state is known as phase space. Theoretically speaking, the phase space of a nonlinear system may have a very high, even infinite dimension. Actually, a sequence is often acquired: $x_1, x_2, \ldots, x_n$. Formula 2.35 stands for a time series outputted from a continuous dynamic system: $x(1), x(2), \ldots, x(M)$, where M refers to the length of the time series. While reconstructing the phase space, appropriate embedding dimension m and time delay τ should be selected.

$$\begin{gathered} X(1) = [x(1), x(1+\tau), x(1+2\tau), \ldots, x(1+(m-1)\tau)], \\ X(2) = [x(2), x(2+\tau), x(2+2\tau), \ldots, x(2+(m-1)\tau)], \\ \cdots\cdots \\ X(M) = [x(M), x(M+\tau), x(M+2\tau), \ldots, x(M+(m-1)\tau)] \end{gathered} \tag{2.35}$$

Evaluating the effects of phase space reconstruction is to verify whether some invariants of the original phase space can remain unchanged. A sequence can be used directly to analyze the system. However, the nonlinear time series is comprehensive reflection of many interacting physical factors and contains the locus of all variants participating in the movement. Therefore, this time series must be expanded into three- or more-dimensional phase space to fully present the time series information. This is phase space reconstruction of the time series.

In 1981, Takens F. put forward a delay embedding theorem which gives the conditions under which a chaotic dynamic system can be reconstructed from a sequence of observations of the state of a dynamic system. This method has become the primary and most fundamental method for phase space reconstruction. For example, as the chaotic property of voice is mainly generated by incentive airflow, linear prediction error signals are equivalent to incentive signals. Linear prediction method eliminates the influence of vocal tract resonance during the process of voicing. The research findings include: (1) Different voice has different chaotic attractor; (2) the attractor of voiced sound has a closed torus while unvoiced sound has an irregular curve, as shown in Fig. 2.19.

2. Takens Reconstruction Theorem

Suppose M is an m-dimensional manifold, for transform pair (ϕ, y), $\phi : M \rightarrow M$ is a smooth diffeomorphism and y is a smooth function in M, then $\Phi_{(\phi,y)} : M \rightarrow R^{2m+1}$ is an embedding. Here, $\Phi_{(\phi,y)} = \left\{y(x), y[\phi(x)], \ldots, y\left[\phi^{2m}(x)\right]\right\}$, where ϕ' is a stream of x, ϕ corresponds to the dynamic relationship of a dynamic system, M to the attractor of the system, and y to the function relationship between the system state and the measurement data. Embed a chaotic time series of a single variant $x_1, x_2, \ldots\ldots x_n$ in m-dimensional space to get the phase space locus for N phase points:

$$Y_i = \left(x_i, x_{i+\tau}, \ldots, x_{i+(m-1)\tau}\right)^T \tag{2.36}$$

In the formula, $i = 1, 2, \ldots, N$; $N = n - (m - 1)\tau$; Y_i stands for the phase space vector after reconstruction; τ for delay time; m for embedding dimension; n for points of the original time series; N for the number of phase space vectors after reconstruction. Therefore, the phase space matrix is acquired:

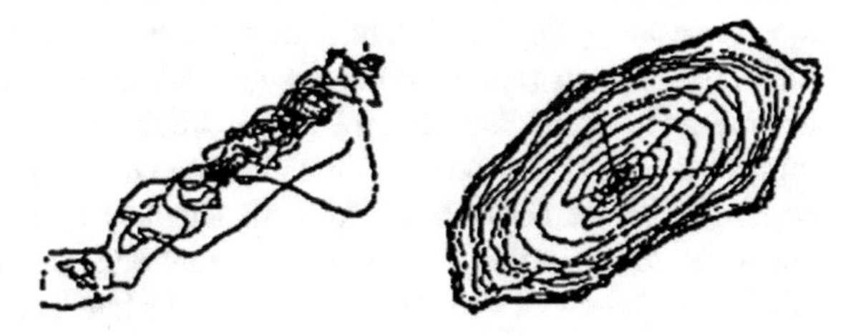

Fig. 2.19 Phase space locus of chaotic attractors for unvoiced and voiced sound

$$\begin{cases} Y_1 = \left(x_1, x_{1+\tau}, \ldots, x_{1+(m-1)\tau}\right)^T \\ Y_2 = \left(x_2, x_{2+\tau}, \ldots, x_{2+(m-1)\tau}\right)^T \\ \cdots\cdots\cdots\cdots\cdots\cdots\cdots\cdots\cdots \\ Y_N = \left(x_N, x_{N+\tau}, \ldots, x_{N+(m-1)\tau}\right)^T \end{cases} \tag{2.37}$$

According to Takens Theorem, with proper τ and m selected, the dynamic form of the original system can be restored with topological equivalence to identify the basic property of the original dynamic system. Moreover, during the evolution process of a dynamic system, all variants are inherently correlated and the evolution of any variant in the system is dependent upon its interaction with other variants in the system. Therefore, evolution information of the system is contained during the evolution process of every variant.

To get geometric structure of the phase space of a dynamic system from a sequence, Packard et al. adopted a time-delaying technology to reconstruct phase space. To reconstruct an equivalent state space, only the change of one variant in the system needs to be considered. Delay phase diagram method is used to process the data with fixed delay as new unidimensional information, i.e., map the data of a single variant on a vector point in multidimensional space. Phase space can be reconstructed from single variants and the vector points in the reconstructed phase space have the same property as those in the original true space.

3. Defining Embedding Dimension of the Phase Space and Selection of Optimal Time Delay

To reconstruct phase space of signals based on chaotic theorem, embedding dimension and delay time should be determined at first. In 1983, to extract feature exponents of time series, Grassberger et al. presented the G-P algorithm to calculate fractal dimension of strange attractors. In 1985, Wolf raised an algorithm to calculate Lyapunov exponent of time series. Phase space reconstruction is an important foundation for exponent calculation. Without effective phase space reconstruction of the original system, the above-mentioned calculation cannot be executed. Packard et al. suggested using the delay coordinate of certain variant in the original system to reconstruct phase space. Takens demonstrated that a proper embedding dimension can be found. That means, if the dimension of the delay coordinate $m \geq 2d + 1$(d stands for the dimension of the original dynamic system), regular locus (attractors) in the embedding space can be recovered. However, according to Takens Theorem, the precondition for reconstruction is to get pure data without length and accuracy limits for certain variant in the original dynamic system, which cannot be satisfied in an actual system. Therefore, research must be conducted to determine how long the actual data should be to meet actual reconstruction requirements. In 1988, Smith stated a harsh condition: up to 42^m (m stands for embedding dimension) data points are needed, which was proven unnecessary by Tsonis. In 1990, Ramsey stated that limited data points can be used to get reliable reconstructed phase space; Nvrengerg and Essex presented the minimal data length necessary for reconstruction, for example, for a system with $d < 4$, at least 4000 data points are needed for reliable reconstruction. In 1993, M.T. Rosenstein et al. put forward a method to use a small amount of data to calculate the maximum Lyapunov exponent. For a discrete dynamic system, this method does need a small amount of data. For Henon mapping, Logistic mapping, etc., only 500 data points (maybe fewer) are needed to accurately figure out the maximum Lyapunov exponent. But for continuous dynamic systems such as Lorenz system and Rossler system, the required amount of data is roughly equivalent to the amount of data stated by Nvrengerg.

While reconstructing phase space, selection of proper time delay τ and embedding dimension m has a direct bearing on the accuracy of variants which describe the feature of strange attractors after phase space reconstruction. For selection of τ and m, there are mainly two viewpoints:

According to the first viewpoint, τ and m are independent of each other, and their selection can be conducted independently (Takens proved that for the time series without length limits and noise interference, τ and m are independent of each other). At present, selection of time delay and embedding dimension is based on three rules: (1) sequence correlation methods such as autocorrelation method, mutual information method, and high order correlation method; (2) phase space expansion methods such as factor filling method, vibration measurement method, average displacement method, and SVF method; (3) complex autocorrelation and unbiased complex autocorrelation method.

According to the second viewpoint, τ and m are correlated because the actual time series has limited length and is subject to the influence of various noises. Numerous experiments demonstrated that the relationship between τ and m is closely related to the time window (tw) for phase space reconstruction. For particular time series, tw is relatively stable. The improper matching of τ with m will have a direct influence upon the equivalent relationship between the structure of reconstructed phase space and the original space. Therefore, the combined algorithms of τ and m, such as time window method, C2C method, and automatic algorithm of embedding dimension and time delay, are generated correspondingly.

Most researchers think the second viewpoint is more practical and rational in engineering practice. At present, research of combined algorithms of embedding dimension and time delay is still one of the hot topics for analysis of chaotic time series.

4. Lyapunov Exponent Based on Phase Space Reconstruction

The basic feature of chaotic movement is the extreme sensitivity of movement to the initial condition. With a tiny change of the initial condition, the evolution locus of the system based on time will separate from the original locus at exponential speed and completely cover the actual state of the system after a certain amount of time, which indicates unpredictability of the long-term performance of the system. For unidimensional dynamic system F:

$$\lambda = \lim_{n \to \infty} \frac{1}{n} \sum_{i=1}^{n} \left| \frac{dF(x)}{dx} \right|_{x=x_i} \tag{2.38}$$

λ refers to Lyapunov exponent, which describes the separation degree of the exponent for each of the multiple iterations on average. When $\lambda \leq 0$, viewed as a whole

$$\left| \frac{dF(x)}{dx} \right| \leq 1 \tag{2.39}$$

Therefore, neighboring points will eventually get close and merge into one point, corresponding to stable fixed point and periodic point; $\lambda > 0$ indicates instability of the movement orbit and the orbit of neighboring points separate in exponential way, thus entering into chaos. The maximum Lyapunov exponent can be used to measure the sensitivity of the chaotic system to the initial value. The larger the exponent value is and the stronger the chaotic property is, the higher the sensitivity is; otherwise, the lower the sensitivity is. Lyapunov exponent indicates the separation level of two neighboring points which evolve with time in the reconstructed phase space. Lyapunov exponent is a pedigree $\lambda_i (i = 1, 2, \ldots, p-1)$, where λ_i indicates the separation level of neighboring orbits which evolve with time in each dimension of the m-dimensional space. If only the maximum exponent in the pedigree is positive, the system is a unidimensional chaotic system; if two or more Lyapunov exponents are positive, the system is a multidimensional chaotic or super chaotic system; or the system is not chaotic. Wolf et al. proposed that the method based on phase locus evolution be used to estimate the maximum Lyapunov exponent $\lambda_{\max}$ with the following steps:

(1) Reconstruct phase space.
(2) With the initial phase point Y_{t_0} as starting point, suppose Y'_{t_0} is the point closest to Y_{t_0}, construct the initial vector V_0, and calculate the length L_{t_0}. Suppose V_0 evolves forward along the locus in a proper period of time τ to get a new vector V_1, calculate the length L'_{t_0}; find $Y'(t_1)$ at $t_1 = t_0 + \tau$, and track and calculate L_{t_1} and L'_{t_1}; repeat the process for m times to get the calculation of maximum Lyapunov exponent λ:

$$\lambda = \frac{1}{N_m} \sum_{k=1}^{M} \ln \frac{L'_{t_k+1}}{L_{t_k}} \tag{2.40}$$

In the formula, N_m stands for the total points of the time series in the m-dimensional space; M for the groups formed by (L_{t_k}, L'_{t_k+1}); L_{t_k} for the distance between the point $X_{t_k} = (x_{t_k}, x_{t_k-1}, \ldots x_{t_k-m+1})$ and the point closest to it; L'_{t_k+1} for the length of L_{t_k} at $t_k + 1$. When the estimated value of the exponent remains stable with m, the acquired calculation result is $\lambda_{\max}$, the maximum Lyapunov exponent.

5. Defining Step Size for Multistep Prediction

As chaotic behavior is resulted from the presence of strange attractors, its short-term behavior is predictable. However, because of its extreme sensitivity to the initial condition, its long-term behavior cannot be predicted. For a time series without noise, the longer it is, the higher the prediction accuracy is and more slowly the prediction accuracy decreases with the step size. When the time series is long enough, higher prediction accuracy can be achieved even for multistep prediction. For the time series with noise, the prediction accuracy decreases at exponential speed with the increase of step size and the length of the time series can hardly influence the prediction time. But the higher the noise level is, the shorter the prediction time is. Many scholars define the maximum prediction time length of the chaotic series as the reciprocal of the maximum Lyapunov exponent λ, or define the average prediction time ($T_{\max} = 1/k$) according to Kolmogorov entropy. Researches indicate that it is inadvisable to define the maximum prediction time length with the reciprocal of the maximum Lyapunov exponent λ without considering the influence of such factors as specific object, series length, and noise level upon the prediction time. Therefore, in this book, study object and series length are taken into consideration and improved maximum prediction time length is adopted after experimental demonstration:

$$T_{\max} = 1/\lambda + 1 \tag{2.41}$$

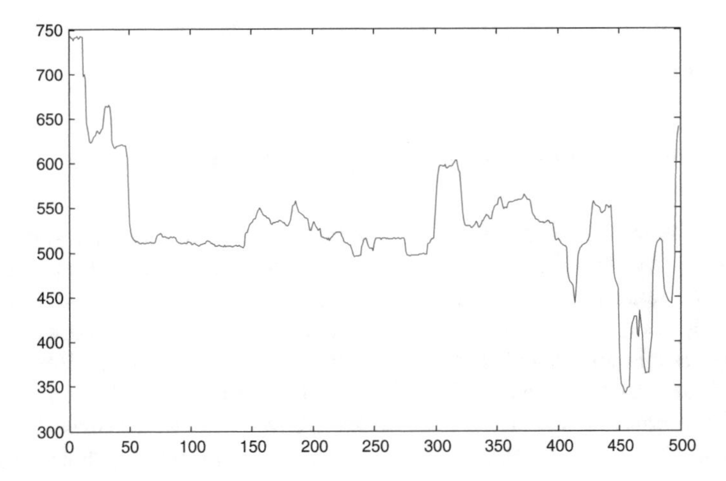

Fig. 2.20 *T*4 temperature

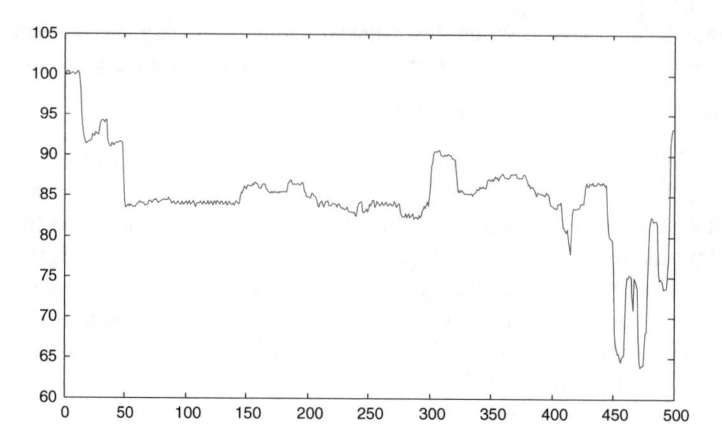

Fig. 2.21 $N1$, low-pressure rotor speed

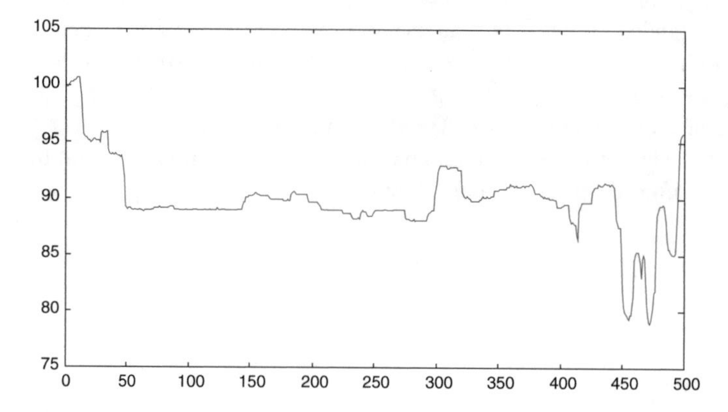

Fig. 2.22 $N2$, high-pressure rotor speed

2.5.2 *Analysis and Verification of Chaotic Property*

To reveal the in-depth change feature of flight data, chaotic analysis theory is used to determine whether the flight data series has chaotic property. At first, define the embedding dimension m and time delay t for phase space reconstruction of flight data $\{x(t), t = 1, 2, \ldots, n\}$; then, use m and t to reconstruct phase space, construct learning sample and calculate the maximum Lyapunov exponent. Here, typical engine state data are adopted, including $T4$ temperature, $N1$ and $N2$ for low-pressure and high-pressure rotor speed, respectively, with series length n as 500, as shown in Figs. 2.20, 2.21, and 2.22, where horizontal coordinates indicate the number of data.

1. Phase Space Reconstruction of Flight Data

As flight data are correlated in various ways, in the analysis of chaotic time series, the first step for chaotic prediction and differentiation of the features of chaotic signals is to reconstruct phase space of the chaotic signals. Therefore, it is necessary to define proper embedding dimension m and delay time τ.

1) Selection of Delay Time

Selection of proper delay time can guarantee the independence and weak correlation of embedding coordinates. One of the relatively simple and practical ways to determine delay time τ is to adopt series autocorrelation function. Make 1–200 s delay for three series with autocorrelation function method. Figures 2.23, 2.24, and 2.25 (the horizontal coordinate axis X refer to delay time(s), the vertical coordinate axis Y refer to autocorrelation function value) show the autocorrelation function curves of delay time τ with temperature $T4$ and high-pressure/low-pressure rotor speed, respectively. The figures show that the autocorrelation function curves heave and set at $s = 0$ with increasing of delay time. In general, the time when the autocorrelation value becomes zero (or approximate to zero) for the first time is selected as delay time to be defined.

As shown in Fig. 2.23, with the autocorrelation functions of $T4$ temperature: $k(1) = 1.0679$, $k(2) = 1.2496$, $k(3) = 2.1963$, and $k(4) = -2.8357$, $k(1)$ is the value when the autocorrelation function of $T4$ temperature is approximate to zero for the first time, and thus the optimal delay time of $T4$ temperature can be determined as $\tau = 1$.

As shown in Fig. 2.24, with the autocorrelation functions of low-pressure speed $N1$: $k(1) = 1.0142$, $k(2) = 1.0917$, $k(3) = 1.2888$, and $k(4) = -5.2480$, $k(1)$ is the value when the autocorrelation function is approximate to zero for the first time, and thus the optimal delay time of $N1$ can be determined as $\tau = 1$.

As shown in Fig. 2.25, with the autocorrelation functions of high-pressure speed $N2$: $k(1) = 1.0202$, $k(2) = 1.0413$, $k(3) = 1.0861$, $k(4) = 1.3471$, $k(5) = 3.7498$, and $k(6) = -1.9008$, $k(1)$ is the value when the autocorrelation function is approximate to zero for the first time, and thus the optimal delay time of $N2$ can be determined as $\tau = 1$.

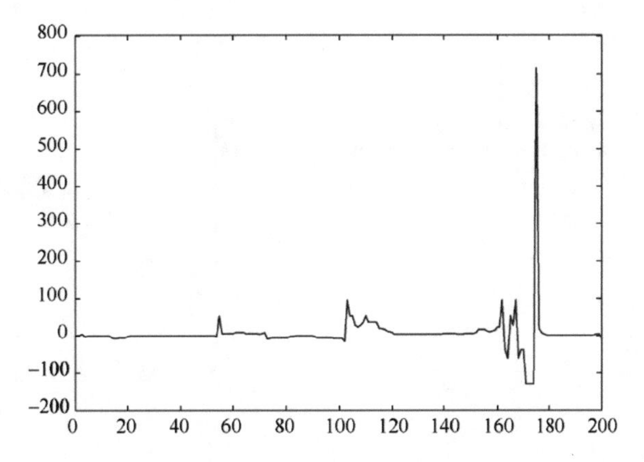

Fig. 2.23 Delay time curve of $T4$ temperature

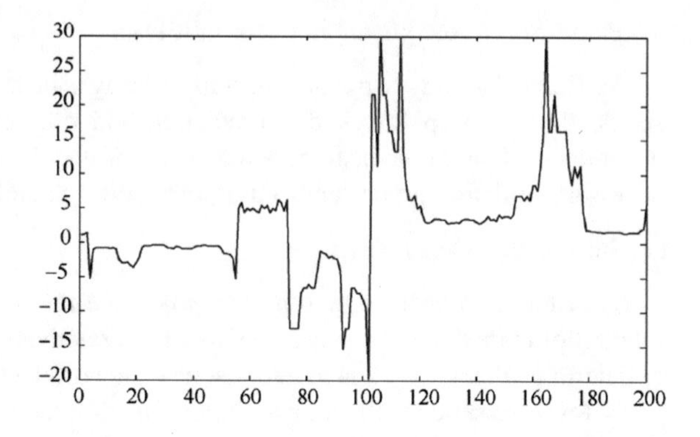

Fig. 2.24 Delay time curve of low-pressure speed $N1$

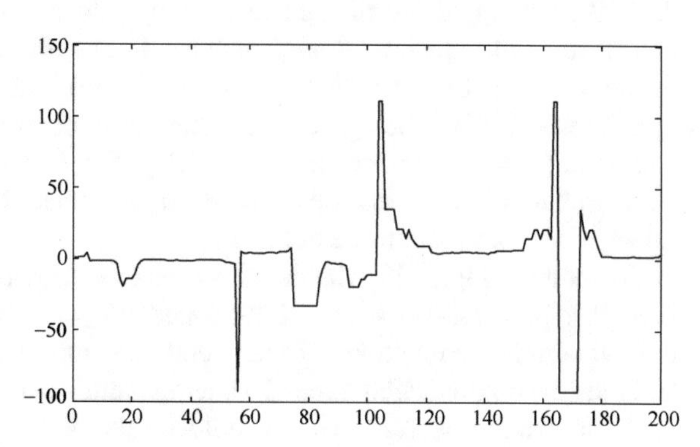

Fig. 2.25 Delay time curve of high-pressure speed $N2$

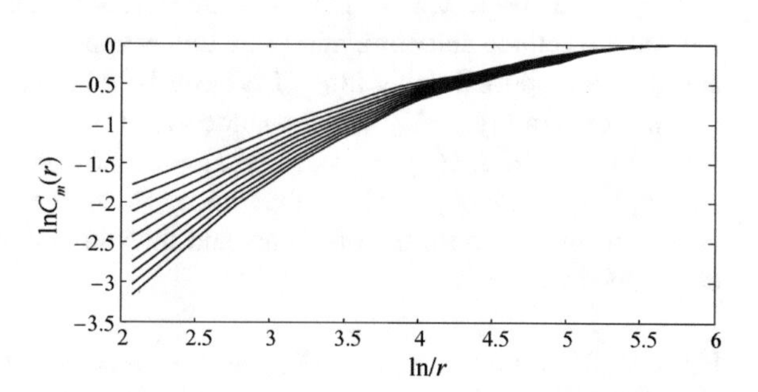

Fig. 2.26 $\ln C_m(r) - \ln r$ curve of $T4$ temperature

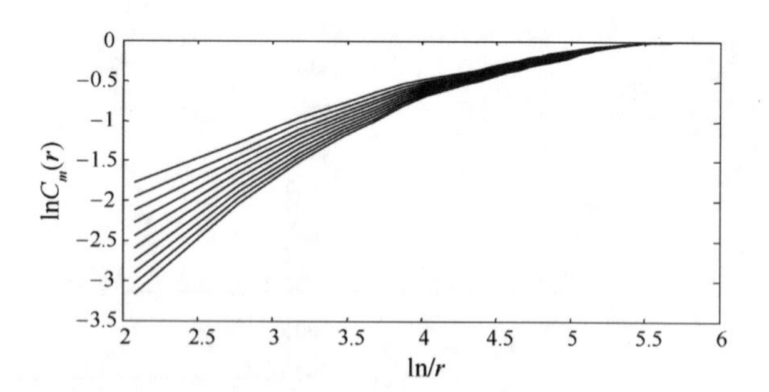

Fig. 2.27 $\ln C_m(r) - \ln r$ curve of low-pressure speed $N1$

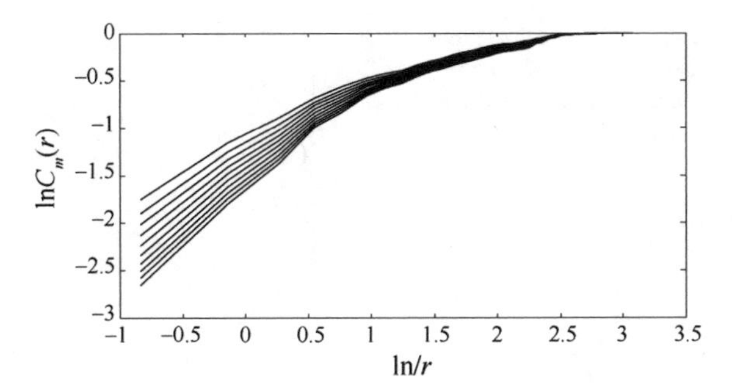

Fig. 2.28 $\ln C_m(r) - \ln r$ curve of high-pressure speed $N2$

2) Selection of Embedding Dimension

Saturated correlation dimension algorithm (briefly known as G-P algorithm) is a common method used to determine embedding dimension. Correlation dimension will converge gradually with the rising of the embedding dimension and eventually result in a convergence value. The dimension in the process of converging is the saturated embedding dimension. According to G-P algorithm, when embedding dimension m is the value from 1 to 10, the $\ln C_m(r) - \ln r$ curves of $T4$ temperature and high-pressure/low-pressure speed are shown in Figs. 2.26, 2.27, and 2.28.

As shown in Figs. 2.26, 2.27, and 2.28, with the rising of embedding dimension m, the line segments in the middle become parallel gradually. Eliminate the line segments with slope as 0 and ∞ and determine the optimum fitting straight line. The slope of the straight line is the correlation dimension. The curves of the correlation dimension D of $T4$ temperature and high-pressure/low-pressure speed changing with the embedding dimension m are shown in Figs. 2.29, 2.30, and 2.31.

As Fig. 2.29 shows, with the rising of the embedding dimension, the correlation dimension of $T4$ temperature with $m = 6$ becomes saturated, and the corresponding correlation dimension $D(6) = 2.6$. As Fig. 2.30 shows, the correlation dimension of low-pressure speed $N1$ becomes saturated when $m = 4$, and the corresponding correlation dimension $D(4) = 1.71$. As Fig. 2.31 shows, the correlation dimension of high-pressure speed $N2$ becomes saturated when $m = 5$, and the corresponding correlation dimension $D(5) = 1.9$. Therefore, 6, 4, and 5 are selected for the embedding dimension of $T4$ temperature and high-pressure/low-pressure speed, respectively.

For selection of saturated correlation dimension, a very large number of samples are required. Smith states that $N_{\min} \geq 42^m$, the minimum sample number is required. Thus, when $m = 3$, $N_{\min} \geq 74088$. This requirement is relatively demanding and hard to be met. Eckman and Ruelle derived $N_{\min} \geq \sqrt{10}^D$, where D stands for the correlation dimension. When $D = 4$, $N_{\min} \geq 100$. This requirement is less demanding. Hong Shizhong and Hong Shiming analyzed the derivation of these two conclusions and found some problems. So, they summarized a new relationship formula, $N_{\min} \geq \sqrt{2}(\sqrt{27.5})^D$. With this method, when $D = 3$, $N_{\min} \geq 204$. Therefore, this method is relatively appropriate. Lin Zhenshan believes that a large

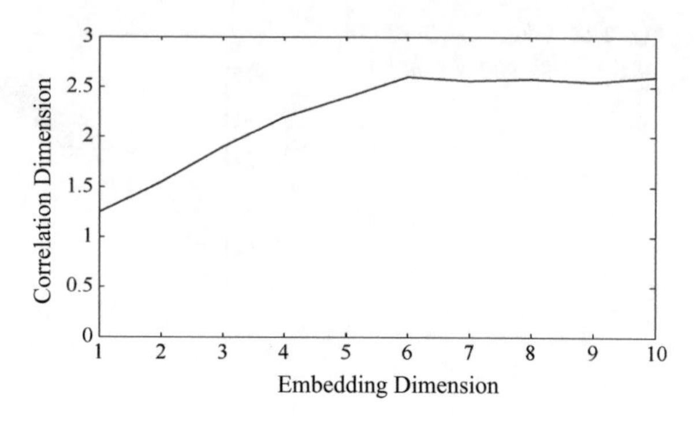

Fig. 2.29 Relationship between the embedding dimension and the correlation dimension of $T4$

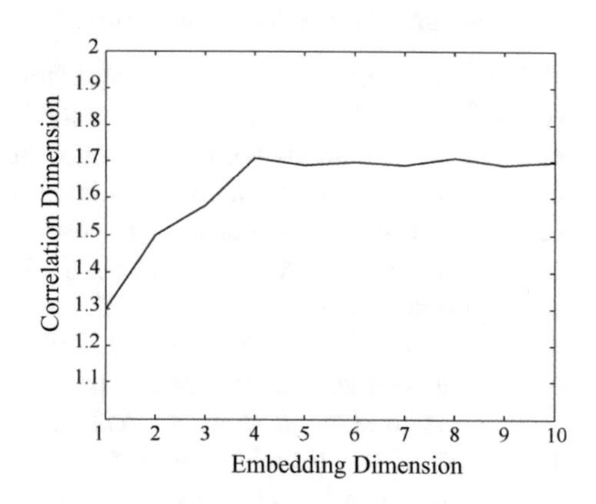

Fig. 2.30 Relationship between the embedding dimension and the correlation dimension of $N1$

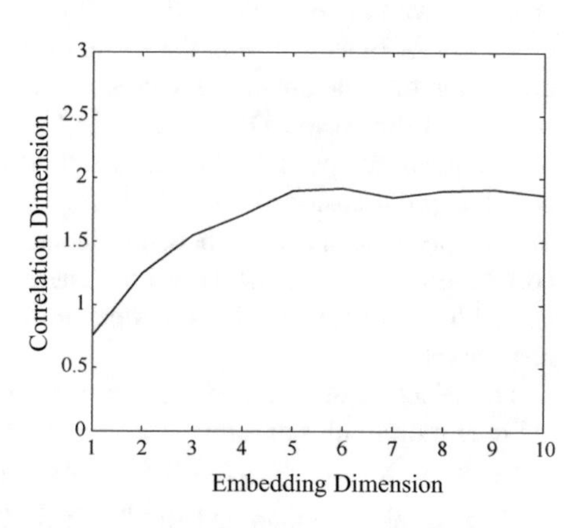

Fig. 2.31 Relationship between the embedding dimension and the correlation dimension of $N2$

amount of data is unnecessary for phase space reconstruction aimed at prediction and only enough data are needed for experiments to realize $N_{\min} \geq 260$. The data adopted in this book are 500, which meets the requirement.

2. Qualitative Analysis of Chaotic Property

Qualitative analysis of chaotic property is to roughly analyze the property of measured series in time or frequency domain. The common methods include phase diagram method, power spectrum, etc.

Phase diagram can describe the changes of the system state during the whole time period and reflect the spatial structure of system attractors. The phase space locus of a chaotic system usually manifests as iterative, aperiodic, and never intersecting movement resulted from its stretching and folding in limited space. It differs from irregular and random movement and it is not a repeated movement of a periodic function.

Power spectrum can be used to differentiate the regular (fixed point, periodic, quasi-periodic) and irregular forms (chaos, noise) of the time series. The power spectrum for periodic movement is discrete, only including basic frequency and its harmonic wave or frequency division. The power spectrum for random white noise and chaos is continuous while the power spectrum for chaotic series is continuous with broad peak. Power spectrum method is adopted to analyze the above data and Figs. 2.32, 2.33, and 2.34 show the power spectrum of $T4$ temperature and high-pressure/low-pressure speed, respectively. As the figures show, the curves are all continuous with broad peak. Therefore, the three series are identified initially as chaotic series.

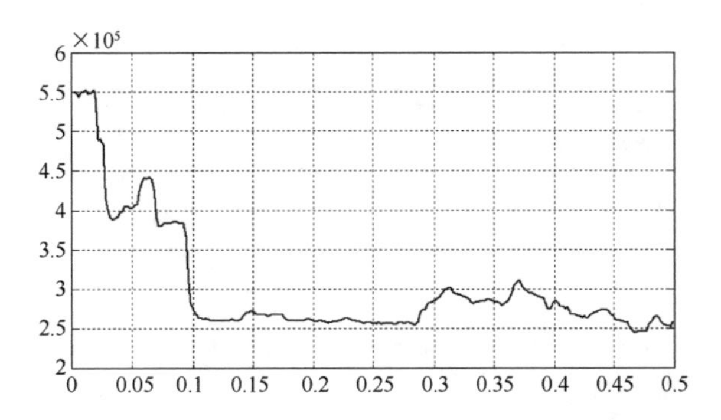

Fig. 2.32 Power spectrum of $T4$ temperature

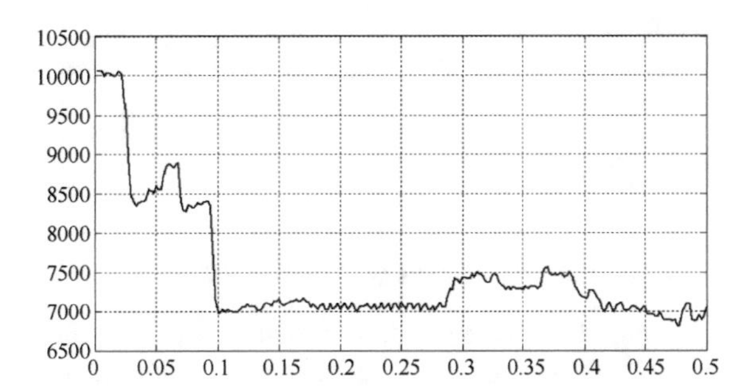

Fig. 2.33 Power spectrum of low-pressure speed $N1$

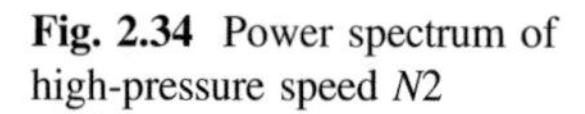

Fig. 2.34 Power spectrum of high-pressure speed $N2$

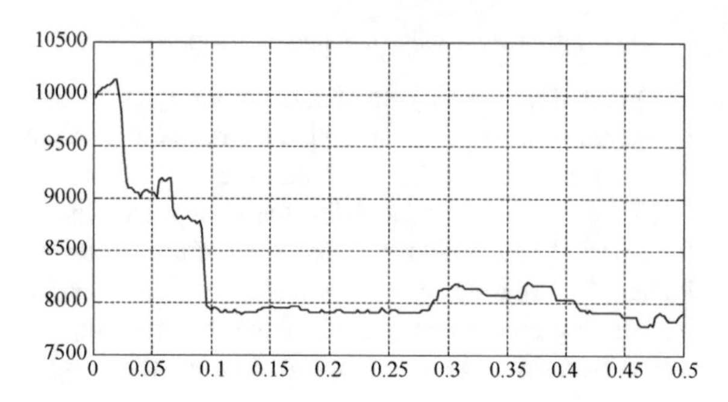

Table 2.3 Maximum Lyapunov exponent for $T4$ temperature

Embedding dimension of $T4$	2	3	4	5	6	7	8
Lyapunov exponent for $T4$ temperature	0.1156	0.1607	0.1231	0.1087	0.1008	0.1411	0.0584

Table 2.4 Maximum Lyapunov exponent for low-pressure speed $N1$

Embedding dimension of low-pressure speed $N1$	2	3	4	5	6	7	8
Lyapunov exponent for low-pressure SPEED $N1$	0.1831	0.1136	0.1178	0.1012	0.0876	0.0994	0.1086

Table 2.5 Maximum Lyapunov exponent for high-pressure speed $N2$

Embedding dimension of high-pressure speed $N2$	2	3	4	5	6	7	8
Lyapunov exponent for high-pressure speed $N2$	0.0514	0.0734	0.0675	0.0408	0.0345	0.0481	0.0326

3. Quantitative Analysis of Chaotic Property

From the above-mentioned qualitative analysis, rough knowledge about the chaotic property of engine data can be acquired. But it is not enough to prove that the series is chaotic. Therefore, it is necessary to use quantitative analysis to verify it. Methods for quantitative analysis include maximum Lyapunov exponent, Kolmogorov Entropy, etc.

Wolf method is adopted to estimate λ, the maximum Lyapunov exponent in the flight data series. To testify the stability of the algorithm and study the influence of different embedding dimensions m upon λ, calculate λ, the maximum Lyapunov exponent with embedding dimension m from 2 to 8, respectively. Tables 2.3, 2.4, and 2.5 describe the results, which show that with embedding dimension m from 2 to 8, the maximum Pyapunov exponents for $T4$ temperature and high-pressure/low-pressure speed are different, but λ is greater than zero. For $T4$ temperature, when the embedding dimension $m = 3$, the maximum Lyapunov exponent $\lambda = 0.1607$. For low-pressure speed $N1$, when the embedding dimension $m = 2$, the maximum Lyapunov exponent $\lambda = 0.1831$. For high-pressure speed $N2$, when the embedding dimension $m = 3$, the maximum Lyapunov exponent $\lambda = 0.0734$. The conclusion is that data for $T4$ temperature and high-pressure/low-pressure speed have obviously chaotic property. Therefore, the methods involved can provide a feasible thought and approach to identify whether other types of flight data have chaotic property.

Chapter 3
Typical Time Series Analysis of Flight Data Based on ARMA Model

Since flight data is a typical time series, time series analysis of flight data processing is a basic and commonly practiced research method. This chapter begins with an introduction to a general modeling method for time series; then an aircraft steady-state parameter prediction method based on AR model is proposed; and finally the method is verified by real flight data.

3.1 Theory of ARMA Model

Time series is a group of chronologically-ordered datasets which take such forms as numerical value, voice, video, text, and web page data. All the flight data time series mentioned and discussed in this book are numerical sequences unless otherwise specified. When the variable $x(t)$ of a physical process is observed and measured, if the discrete ordered set $\boldsymbol{X} = \{x_{t_1}, x_{t_2}, \ldots, x_{t_n}\}$ is obtained at the time $t_1, t_2, \ldots, t_n (t_1 < t_2 < \cdots < t_n)$, this discrete ordered set is a discrete numerical time series. Suppose $x(t)$ is a stochastic process, then $x_{t_i} (i = 1, 2, \ldots, n)$ is a sample implementation, i.e. a time series. The amount of sampling for time series x is called sequence length N, expressed as $N = \mathrm{len}(x)$. Time series analysis is a data processing method which analyzes and processes the observed random data by means of parameter models. It focuses on the study of the laws governing the inherent relationship of the ordered random data sequences, without the need to know in advance the system inputs and the property of its dynamitic process. It studies system property on the basis of the intrinsic laws of the output data sequences. Since Auto-Regressive Moving Average model (ARMA model) is the most typical

J. Zhang and P. Zhang, *Time Series Analysis Methods and Applications for Flight Data*, DOI 10.1007/978-3-662-53430-4_3

and widely used time series model, a basic introduction to this model is necessary in this section.

3.1.1 Mathematical Model

For $\{x_t\}$, a stable time series with mean value as zero, $t = 1, 2, \ldots, n$, a fitted stochastic difference equation can be given as follows:

$$x_t - \phi_1 x_{t-1} - \phi_2 x_{t-2} - \cdots - \phi_n x_{t-n} = a_t - \theta_1 a_{t-1} - \theta_2 a_{t-2} - \cdots - \theta_m a_{t-m} \tag{3.1}$$

or

$$x_t = \sum_{t=1}^{n} \varphi_i x_{t-i} - \sum_{j=1}^{m} \theta_j a_{t-j} + a_t, a_t \sim \text{NID}\left(0, \sigma_a^2\right) \tag{3.2}$$

The above formula represents an n-order autoregressive, and m-order moving average model, i.e. $\text{ARMA}(n, m)$, in which x_t is an element of time series $\{x_t\}$ at time t; $\phi_i\,(i = 1, 2, \ldots, n)$ is autoregressive coefficient; $\theta_j\,(j = 1, 2, \ldots, m)$ is moving average coefficient; a_t is residual error, which is white noise input at time i, a mutually independent variable in normal distribution.

When $\theta_j = 0$, $\text{ARMA}(n, m)$ is reduced to $\text{AR}(n)$, an n-order autoregressive model expressed as:

$$x_t = \sum_{t=1}^{n} \varphi_i x_{t-i} + a_t, a_t \sim \text{NID}\left(0, \sigma_a^2\right) \tag{3.3}$$

When $\phi_i = 0$, $\text{ARMA}(n, m)$ is reduced to $\text{MA}(m)$, an m-order moving average model expressed as:

$$x_t = a_t - \sum_{j=1}^{m} \theta_j a_{t-j} + a_t, a_t \sim \text{NID}\left(0, \sigma_a^2\right) \tag{3.4}$$

When operator B (backward shift operator) is introduced into $\text{ARMA}(n, m)$ model, and let

$$\varphi(B) = \left(1 - \sum_{i=1}^{n} \varphi_i B^i\right), \quad \theta(B) = \left(1 - \sum_{j=1}^{n} \theta_j B^j\right)$$

then $\text{ARMA}(n, m)$ can be re-expressed as

$$\phi(B)x_t = \theta(B)a_t, a_t \sim \mathrm{NID}(0, \sigma_a^2) \tag{3.5}$$

The ARMA model comes from a deep physical background and has profound physical meanings. In the mathematical-statistics perspective, ARMA is a converter which converts related stable time series into independent stable ones. In the signal processing perspective, ARMA is an estimator which makes an estimation of the unknown data by using the data available. If the unknown data is historical, ARMA serves as a smoother; but for the unknown future data, ARMA is a predicator. In the information theory perspective, ARMA is an information agglomerator which agglomerates information extracted from massive data into several model parameters for easy analysis and processing. Finally, in the system theory perspective, ARMA is a discrete dynamic system equation with white noise as its input, and an output equivalent to that of a practical physical system. This equation is, therefore, capable of reflecting the features of the system.

3.1.2 Modeling Process

1. Data Acquisition

Two main problems should be considered in collecting data for ARMA model: sampling frequency and sample length. Firstly, ARMA is designed for analysis of discrete time series in which discrete sampling must be done for continuous signals obtained. Secondly, even when system output signals are discrete time series, it is not necessary to include for analysis all the data whose length is within the sample length limit. Hence, sampling frequency and sample length must be defined in advance in order to obtain intended information from the signals.

1) Defining of Sampling Angular Frequency ω_s

Sampling angular frequency is defined according to Shannon Sampling Theorem as $\omega_s \geq 2\omega_{\max}$.

2) Defining of Sample Length L

Defining of sample length is mainly related to energy leakage effect of signals in frequency domain and resolutions of harmonics of different frequencies. This is caused by the fact that sample length is limited, that is, only part of the sample is captured for signal windowing processing.

In signal processing, window functions of different shapes are used for energy leakage problem, whereas width L of window function is for frequency resolution problem. Generally speaking,

$$L > 1/(f_2 - f_1) \quad \text{or} \quad N > \omega_s/(f_2 - f_1) \tag{3.6}$$

In the formula, N is the length of the sample, f_1,f_2 are the frequencies of harmonics for two adjacent frequencies in the signal. In practice, the length of the sample can be assigned smaller values of L or N, even as small as $(1/3 \sim 1/4)/L$.

2. Validation of Data Features

ARMA model requires $\{x_t\}$ to be stable and normal time series with mean value as zero, thus whether $\{x_t\}$ satisfying these three conditions should be tested.

1) Stability Test

A stable time series is provided with two basic features: its mean value μ_j and variance σ_j^2 are constant, and auto-covariance function R_k is time independent which is only related to time interval. The following three methods can be used for this stability test.

(1) Segmentation test (parametric test). When sample length of time series $\{x_t\}$ is relatively large, $\{x_t\}$ should be divided equally into l subsequences, $\{x_{1t}\}, \{x_{2t}\}, \ldots, \{x_{lt}\}$ and the length of each subsequence is M, and $N = lM$. For the subsequence $j\{x_{jt}\}$, we can work out the estimated value of its mean value, variance and auto-covariance:

$$\mu_j = \frac{1}{M}\sum_{t=1}^{M} x_{jt}, \quad \sigma_j^2 = \frac{1}{M}\sum_{t=1}^{M} (x_{jt} - \mu_j)^2 \tag{3.7}$$

$$R_{j,k} = \frac{1}{(M-k)}\sum_{t=k+1}^{M} (x_{jt} - \mu_j)(x_{j,t-k} - \mu_j) \tag{3.8}$$

If $\{x_t\}$ is a stable time series, no significant difference should exist for the calculated mean value, variance and auto-covariance function value. Method to test significant difference is as follows.

Take significant level as 0.05, if statistic features of any two subsequences $\{x_{it}\}$ and $\{x_{jt}\}$ satisfy the following relations:

$$\begin{aligned} &|\mu_{\mathrm{i}} - \mu_j| > 2.77\sigma(\mu_j) \\ &\left|\sigma_{\mathrm{i}}^2 - \sigma_{\mathrm{j}}^2\right| > 2.77\sigma(\sigma_{\mathrm{j}}^2), \quad (i \neq j; i,j = 1,2,\ldots,l) \\ &|R_{i,k} - R_{j,k}| > 2.77\sigma(R_{j,k}) \end{aligned}$$

then the differences between u_i and u_j, σ_i^2 and σ_j^2, $R_{i,k}$ and $R_{j,k}$ are significant, hence, $\{x_t\}$ is unstable.

$$\sigma^2(\mu_j) = \frac{\sigma_x^2}{M}\left[1 + 2\sum_{t=1}^{M}\left(1 - \frac{t}{M}\right)R_k\right]$$

$$\sigma^2(\sigma_j^2) = \frac{2\sigma_x^4}{M}\left[1 + 2\sum_{t=1}^{M}\left(1 + \frac{t}{M}\right)R_k^2\right]$$

where $\sigma(\mu_j)$, $\sigma(\sigma_j^2)$ and $\sigma(R_{j,k})$ are theoretical mean square deviations of μ_j, σ_j^2 and $R_{j,k}$ respectively, which can be calculated as

$$\sigma^2(R_{j,k}) = \frac{1}{M-k}\left[1 + R_k^2 + 2\sum_{t=1}^{M-k}\left(1 - \frac{t}{M-k}\right) \times \left(R_t^2 + R_{t+k}R_{t-k}\right)\right] \qquad (3.9)$$

where σ_x^2 is variance of $\{x_t\}$, and R_k is auto-covariance function with its calculation formula as

$$\widehat{R}_k = \frac{1}{N}\sum_{t=k+1}^{N} x_t x_{t-k}, \quad (k = 0, 1, 2, \ldots N-1)$$

$$\widehat{\sigma}_x^2 = \frac{1}{N}\sum_{t=1}^{N} x_t^2$$

(2) Reversed Order Test (non-parametric test). Reversed order test is used to test the mean values and significances of variance differences for the sub-sequences. All mean values compose a sequence $\mu_1, \mu_2, \ldots, \mu_l$ after mean value of each sub-sequence is acquired. When $i > j\,(j = 1, 2, \ldots, l-1)$, there is $\mu_i > \mu_j$ defined as a reversed order of μ_j. The amount of reversed orders A_j is equivalent to the number of occurrences of $\mu_i > \mu_j\,(i > j)$. The sum of all the reversed orders A_j is called the total amount of reversed orders A with its theoretical mean value and variance being described as:

$$A = \sum_{j=1}^{l-1} A_j$$

$$E[A] = \sum_{j=1}^{l-1} E[A_j] = \frac{l(l-1)}{4}$$

$$\sigma_A^2 = \frac{l(2l^2 + 3l - 5)}{72}$$

$$u = \frac{(A + 1/2 - E[A])}{\sigma_A}$$

Construct standard normal distribution statistics u and take significant level as 0.05. If $|u| \leq 1.96$, there is no significant difference among μ_i, hence $\{x_t\}$ is stable time series.

(3) Auto-Covariance Function Test (non-parametric test). According to the relations of Green function G_j and auto-covariance function R_k, when $\{x_t\}$ is a stable time series, then

$$\lim_{k\to\infty} R_k = \lim_{k\to\infty}\left[\sigma_a^2 \sum_{j=0}^{\infty} G_j G_{j+k}\right] = \sigma_a^2 \sum_{j=0}^{\infty} G_j G_{\infty} = 0 \tag{3.10}$$

Hence, $\{x_t\}$ is considered as stable time series when R_k attenuates toward 0 as k increases.

2) Normality Test

It is essential to test whether the third-order moments (skewness coefficient) and the fourth-order moments (kutosis coefficient) of $\{x_t\}$ satisfy the feature requirements of normal random variable. However, as most of the engineering problems possess the feature of normal distribution, it does not necessarily require $\{x_t\}$ to be a normal time series if only construction of ARMA model is needed. Hence, this normality test can be omitted.

3) Zero Mean Value Test

Zero mean value test is used to test whether the truth value μ_x of $\{x_t\}$ is 0 or not. When $\mu_x \neq 0$ and its value is unknown, assuming $\mu_x = \widehat{\mu}_x$, the time series would be annihilated as $E[y_t] = E[x_t - \widehat{\mu}_x] = 0$, where

$$\widehat{\mu}_x = \left(\sum_{t=1}^{N} x_t\right)/N$$

3.1.3 *Estimation of Model Parameters*

As for AR(n) model:

$$x_t = \varphi_1 x_{t-1} + \varphi_2 x_{t-2} + \cdots + \varphi_n x_{t-n} + a_t \tag{3.11}$$

Parameter estimation refers to estimation of n parameters $\varphi_i\ (i = 1, 2, \ldots, n)$. Methods often used for this estimation include Least Square Method and U-C algorithm.

1. Least Square Estimation Method

AR model is derived on the basis of multiple regression models. By substituting time series $\{x_t\}$ into (3.11), the following linear equation set can be obtained:

$$\begin{cases} x_{n+1} = \varphi_1 x_n + \varphi_2 x_{n-1} + \cdots + \varphi_n x_1 + a_{n+1} \\ x_{n+2} = \varphi_1 x_{n+1} + \varphi_2 x_n + \cdots + \varphi_n x_2 + a_{n+2} \\ \cdots \\ x_N = \varphi_1 x_{N-1} + \varphi_2 x_{N-2} + \cdots + \varphi_n x_{N-n} + a_N \end{cases} \tag{3.12}$$

The equation set can be described in matrix form as

$$\boldsymbol{Y} = \boldsymbol{X}\boldsymbol{\varphi} + \boldsymbol{a} \tag{3.13}$$

where

$$\begin{cases} \boldsymbol{Y} = [x_{n+1}, x_{n+2}, \ldots, x_N]^T \\ \boldsymbol{\varphi} = [\varphi_1, \varphi_2, \ldots, \varphi_n]^T \\ \boldsymbol{a} = [a_{n+1}, a_{n+2}, \ldots, a_N] \\ \boldsymbol{X} = \begin{bmatrix} x_n & x_{n-1} & \cdots & x_1 \\ x_{n+1} & x_n & \cdots & x_2 \\ \vdots & \vdots & \ddots & \vdots \\ x_{N-1} & x_{N-2} & \cdots & x_{N-n} \end{bmatrix} \end{cases} \tag{3.14}$$

According to multiple regression theory, the least square estimation of parameter matrix φ is

$$\widehat{\boldsymbol{\varphi}} = (\boldsymbol{X}^T\boldsymbol{X})^{-1}\boldsymbol{X}^T\boldsymbol{Y} \tag{3.15}$$

2. U-C Algorithm

This algorithm was developed by T.J. Ulrych and R.W. Clayton in 1976. According to U-C algorithm, the least square estimation uses the observed time series $\{x_t\}$ only once, without making the utmost of information contained in $\{x_t\}$. If time series $\{x_t\}$ can be used several times in the process of parameter estimation, all the information in it can be utilized fully, thus precision of parameter estimation would be improved. Therefore, U-C algorithm could align the observed time series $\{x_t\}$ positively (positive alignment) and reversely (reverse alignment).

Positive alignment for $x_1, x_2, \ldots, x_N$ is

$$\{x_t\} \quad (t = 1, 2, \ldots, N)$$

Reverse alignment for $x_N, x_{N-1}, \ldots, x_1$ is

$$\{x_t\} \quad (t = N, N-1, \ldots, 1)$$

Theoretically speaking, if $\{x_t\}$ is stable, information from positive and reverse time series should be consistent. However, $\{x_t\}$ is just a sample with limited length, and it is impossible in practice for the time series to be always stable. Hence, information provided by positive and reverse time series is not consistent. Therefore, positive and reverse alignments of time series are jointly used in U-C algorithm to estimate parameters.

According to signal processing theory, the item $\varphi_1 x_{t-1} + \varphi_2 x_{t-2} + \cdots + \varphi_n x_{t-n}$ in AR model is called filtering value and a_t is filtering error. Hence, forward filtering error is assumed as $f_{n,t} = a_t$ with its value as

$$f_{n,t} = x_t - \varphi_1 x_{t-1} - \varphi_2 x_{t-2} - \cdots - \varphi_n x_{t-n} \tag{3.16}$$

The first subscript of $f_{n,t}$ refers to the order of the AR model as n, then the corresponding AR model can be expressed as

$$x_t = \phi_1 x_{t-1} + \phi_2 x_{t-2} + \cdots + \phi_n x_{t-n} + f_{n,t} \tag{3.17}$$

By substituting the positive time series $\{x_t\}$ into the above formula, linear formula set which is equivalent to the above formula can be obtained, which is expressed as

$$\boldsymbol{Y}_f = \boldsymbol{X}_f \boldsymbol{\varphi} + \boldsymbol{f} \tag{3.18}$$

In the formula, $\boldsymbol{Y}_f$, $\boldsymbol{X}_f$, $\boldsymbol{f}$ are the same as $\boldsymbol{Y}$, $\boldsymbol{X}$, $\boldsymbol{a}$. Different symbols are used here just for convenience of the following description. They are matrixes respectively composed of positive time series.

If AR model is constructed with reverse time series $\{x_t\}$, it can be expressed as

$$x_{t-n} = \varphi_1 x_{t-n+1} + \varphi_2 x_{t-n+2} + \cdots + \varphi_n x_t + a_{t-n} \tag{3.19}$$

where a_{t-n} is assumed as backward filtering error $b_{n,t}$ with its value as

$$b_{n,t} = x_{t-n} - \varphi_1 x_{t-n+1} - \varphi_2 x_{t-n+2} - \cdots - \varphi_n x_t \tag{3.20}$$

By substituting the reverse time series $\{x_t\}$ into the above formula, the obtained matrix formula can be expressed as

$$\boldsymbol{Y}_b = \boldsymbol{X}_b \boldsymbol{\varphi} + \boldsymbol{b} \tag{3.21}$$

in which

$$\boldsymbol{Y}_b = [x_{N-n}, x_{N-n-1}, \ldots, x_1]^T$$

$$\boldsymbol{b} = \left[b_{n,N-n}, b_{n,N-n-1}, \ldots, b_{n,1}\right]^T$$

$$\boldsymbol{X}_b = \begin{bmatrix} x_{N-n+1} & x_{N-n+2} & \cdots & x_N \\ x_{N-n} & x_{N-n+1} & \cdots & x_{N-1} \\ \vdots & \vdots & \ddots & \vdots \\ x_2 & x_3 & \cdots & x_{n+1} \end{bmatrix}$$

According to least square estimation, the formula (3.21) can be expressed as

$$\begin{cases} \boldsymbol{X}_f^T \boldsymbol{X}_f \boldsymbol{\varphi} = \boldsymbol{X}_f^T \boldsymbol{Y}_f \\ \boldsymbol{X}_b^T \boldsymbol{X}_b \boldsymbol{\varphi} = \boldsymbol{X}_b^T \boldsymbol{Y}_b \end{cases} \tag{3.22}$$

By adding the above two formulas, the following formula is obtained:

$$\left(\boldsymbol{X}_f^T \boldsymbol{X}_f + \boldsymbol{X}_b^T \boldsymbol{X}_b\right) = \boldsymbol{X}_f^T \boldsymbol{Y}_f + \boldsymbol{X}_b^T \boldsymbol{Y}_b \tag{3.23}$$

The formula is called U-C formula, where the matrixes $\boldsymbol{X}_f^T \boldsymbol{X}_f$ and $\boldsymbol{X}_b^T \boldsymbol{X}_b$ are always positive definite square matrixes. Hence inverse matrix exists, with φ estimated as

$$\hat{\boldsymbol{\varphi}} = \left(\boldsymbol{X}_f^T \boldsymbol{X}_f + \boldsymbol{X}_b^T \boldsymbol{X}_b\right)^{-1} \left(\boldsymbol{X}_f^T \boldsymbol{Y}_f + \boldsymbol{X}_b^T \boldsymbol{Y}_b\right) \tag{3.24}$$

3.1.4 Model Applicability Test

The most essential test criterion for model applicability is to test whether $\{a_t\}$ is white noise or not. Based on different experiences, there are four types of criteria: ① test criteria of white noise, ② test criteria of residual sum of squares, ③ Akaike Information Criteria, ④ criteria for special purposes.

The most popular Akaike Information Criteria was developed by Akaike H, with its main idea as: the model's tracing of the real system will be improved in its precision with the model's order increasing, which is presented as the decreasing of residual sum of squares. This is an advantage. However, its disadvantage is that the higher the model's order is, the more model parameters are, thus leading to greater calculation error. Hence, there should be a proper order with its model as the applicable model. Akaike Information Criteria includes the following three types of criteria.

1. FPE Criterion

FPE Criterion is the final prediction error criterion which is suitable for AR(n) model. Take the model with minimum FPE(n) as the applicable model.

$$\mathrm{FPE}(n) = \frac{N+n}{N-n}\sigma_a^2 \tag{3.25}$$

2. AIC Criterion

This criterion starts with the extraction of maximum information from the observed time series, which is well applicable for ARMA(n, m) model. Take the model order p with minimum AIC(p) as the applicable model order.

$$\mathrm{AIC}(p) = N \ln \sigma_a^2 + 2p \tag{3.26}$$

3. BIC Criterion

Some improvements are made to AIC Criterion, which generates $\mathrm{BIC}(p) = N \ln \sigma_a^2 + p \ln N$. The applicable model order determined by BIC Criterion is asymptotical consistent estimation of its truth value.

3.1.5 Optimum Prediction

1. Principle of Optimum Prediction

$\{x_t\}$ can always be expressed by linear combination of a series of white noise $\{a_t\}$, whose weight function is Green Function G_j:

$$x_t = \sum_{j=0}^{\infty} G_j a_{t-j} \tag{3.27}$$

which can be expanded and expressed as

$$x_{t+l} = e_t(l) + \widehat{x}_t(l)$$

In the formula,

$$e_t(l) = \sum_{j=0}^{l-1} G_j a_{t+l-j}\, \widehat{x}(l) = \sum_{j=1}^{\infty} G_j a_{t+l-j}$$

in which $\widehat{x}_t(l)$ and $e_t(l)$ are respectively predictive value and prediction error of the lth step forward, with this error being minimum. Variance of the prediction error is

$$\mathrm{Var}[e_t(l)] = \sigma_a^2 \sum_{j=0}^{l-1} G_j^2$$

Hence, the above prediction is optimum prediction by minimizing the variance of prediction error, which is characterized by the facts that ① the long-term predictive value is the mean value of the time series, and ② variance of long-term prediction error equals variance of the time series itself.

2. Optimum Prediction Calculation of AR Model

The optimum prediction formula for AR(n) model is

$$\hat{x}_t(l) = \begin{cases} \sum_{i=1}^{n} \varphi_i x_{t+l-i} & (l = 1) \\ \sum_{i=1}^{l-1} \varphi_i \hat{x}_t(l-i) + \sum_{i=1}^{n} \varphi_i x_{t+l-i} & (1 < l \le n) \\ \sum_{i=1}^{n} \varphi_i \hat{x}_t(l-i) & (l > n) \end{cases} \tag{3.28}$$

This formula indicates that recursive calculation of $\hat{x}_t(l)$ only uses the n data such as $x_t, x_{t-1}, \ldots, x_{t+1-n}$ and does not involve $\{a_t\}$, which makes the calculation simple.

3. Optimum Prediction Correction

Optimum prediction correction formula is

$$\hat{x}_{t+1}(l-1) = \hat{x}_t(l) + G_{l-1}[x_{t+1} - \hat{x}_t(1)] \tag{3.29}$$

This formula indicates that prediction correction $\hat{x}_{t+1}(l-1)$ for x_{t+l} is composed of two parts: prediction $\hat{x}_t(l)$ at previous time t, and correction for $\hat{x}_t(l)$ with weight of correction term as G_{l-1}.

3.2 Trend of Parameter Monitoring Method Based on AR Model

Traditional time series analysis combines directly parameter model with system analysis, which is not only a data processing method, but also a system research method. According to parameters of a steady-state aircraft, this section introduces a trend of parameter monitoring method based on AR model.

3.2.1 Description for Steady State of Aircraft

Trend monitoring of parameters is one of the important methods for aircraft maintenance, state monitoring and prediction. It is a huge workload to use directly flight data provided by the flight data recorder and the model is too complex to reflect the time domain features of the aircraft's whole performance and statistical information of flight performance. Therefore, trend monitoring and prediction of parameters should be conducted when the aircraft is in a steady state.

The steady state of an aircraft can be understood from the following three aspects: ① corresponding to different operating states of the engines, such as idle, rating, full throttle and afterburning. ② the states determined by satisfying certain filtering conditions, for example, "nozzle contracting or expanding" state. ③ corresponding to different flight phases including take-off, climbing, cruise, landing and so on.

The filtering conditions for different steady states are sets of multi-conditional logic judgment inference rules which are composed of a group of "IF…THEN…" statements. These inference rules should fully represent the specific knowledge of related fields, and are easy to understand, implement, maintain and manage. Specifically speaking, flight data-based reference rules are represented in Table 3.1.

Take a certain aircraft for example. Table 3.2 shows parts of the steady states and their filtering conditions.

3.2.2 Extraction of Monitoring Parameters Based on Rule Reasoning Machine

Rule reasoning machine is the key to determine the state or filtering conditions, which can realize reasoning and output of the filtering conditions. Rule reasoning machine includes rule reasoning mechanism and establishment of rule reasoning database. The process of reasoning is shown in Fig. 3.1, which includes four parts: rule writing, compiling, matching and state output.

1. Rule Writing

Friendly human-machine interface ensures convenience of users to input, maintain and manage the rules according to grammatical requirements.

Table 3.1 Representation of flight data-based reference rules

	Rule No.	Conditions of rules (filtering conditions)	Events (filtering object)
Examples	001	($N1$ < 40 and $T4$ > 350) or ($N1$ > 40 and time > 10)	State of startup
Meaning	For rule No. 1, if "$N1$ is less than 40% and $T4$ is greater than 350 °C", or "$N1$ is more than 40% and lasts for more than 10 s", "state of startup" is tenable		

Table 3.2 Parts of the steady states for the aircraft and their filtering conditions

Filtering object	Filtering conditions
Startup	(*N*1 < 40 and *T*4 > 350) or (*N*1 > 40 and time > 10)
Idle	Beginning: (Ys < 18 or *N*1 < 40) and (*N*1 > 20 and *N*2 > 20) End: Ys > 18 or *N*1 > 40
Full throttle	Beginning: Ys > 58 and (100 < *V* < 200) Normal data recording: *N*1 > 40 and *N*2 > 40 and *T*4 > 400 End: Condition 1, Ys < 58 and time > 4 Condition 2, Ys > 60
Fuel consumption at full throttle	Beginning:Ys > 58 and (100 < *V* < 200) Normal data recording:*N*1 > 40 and *N*2 > 40 and *T*4 > 400 End: (Gh0–Gh1) > 3 and time = 7
Nozzle contracting	Ys < 60 and Ksp = 1 and Ksp_old! = Ksp_now
Nozzle expanding	Ys < 60 and Ksp = 0 and Ksp_old! = Ksp_now

Refer to Table 3.3 for parameter symbols and names

Table 3.3 Reference table for parameter symbols and names

No.	Parameter symbols	Parameter names (unit)	No.	Parameter symbols	Parameter names (unit)
1	*N*1	Low pressure rotor speed (%)	5	Ksp	Indication signal when nozzle is contracting (none)
2	*N*2	High pressure rotor speed (%)	6	*V*	Flight velocity (km/h)
3	*T*4	Engine exhaust gas temperature (°C)	7	Gh0, GH1	Fuel consumption in a unit time interval (*L*)
4	Ys	Throttle lever displacement (°)	8	time	Relative duration (s)

2. Rule Compiling

The essence of rule compiling is to convert the rules expressed by natural language into binary code files which can be understood by computer. In the process of compiling, the system can indicate the type and position of errors if they exist, which can be utilized by users to figure out the sources of errors and make additional rule modifications. Rule compiling involves data type, operators and statements.

(1) Data type

Flight data include continuous and discrete parameters, each of which corresponds to different type of data. In line with the principle of reducing storage space and improving efficiency, different parameters should be able to be endowed with appropriate data types such as float and integer data.

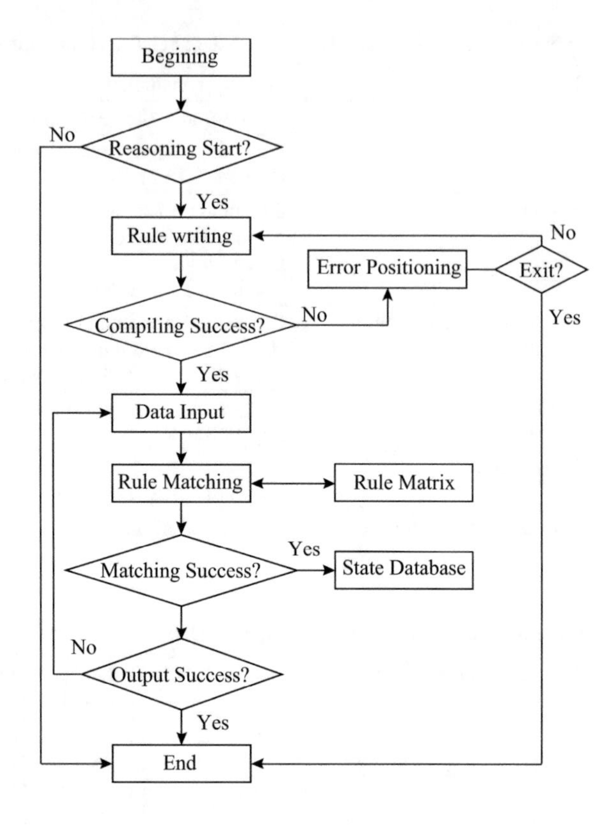

Fig. 3.1 Process of reasoning machine

Table 3.4 Information of operators

Operators	Physical meaning
()	Bracket
Sqrt ()	Evolution
Δ	Variable quantity of parameter in a second
Abs ()	Absolute value
Max(), Min()	Maximum, Minimum
sin(), cos()	sine function, cosine function
*, /	Multiplication, Division
+, −	Addition, Subtraction
&, ||, !	AND, OR, NOT
<, > , <= , >=, ==	Greater than, less than, less than or equal to, greater than or equal to, equal to
=	Assignment

(2) Operators

In order to enable the computer to translate rules, the reasoning machine should support different types of operators including operation function, and execute gradual scanning according to computational priority. Gradual scanning can be described as the process in which computations of different levels should be carried out separately according to the computational priority from high to low, and the computational sequence within a level is defined to be executed from left to right.

Supported operators are suggested according to computational levels from high to low, as shown in Table 3.4.

(3) Statements

Statements are separated, analyzed and recombined according to grammatical rules to determine whether the program of all word symbol strings is grammatically correct. If there are errors which do not conform to grammatical rules, grammar analyzer would inform users of the kinds and positions of the errors so that they can fix some of the bugs.

3. Rule matching

The main task of matching is to simulate thinking process of domain experts to control and execute problem solutions. The basic process is to read in flight data frame by frame and make judgments successively about whether the preconditions of each rule in the rule database are satisfied. If preconditions are satisfied, the results would be output into a specified state database. Hence one rule matching succeeds. If preconditions are not satisfied, the next frame of flight data would be read in until all data processing is finished.

4. State output

State output means rule reasoning is finished. Forms of state output are shown in Table 3.5.

Monitorable parameters are defined based on two aspects. On the one hand, it is dependent on the features of the aircraft subsystems. On the other hand, it is limited by the ability of flight data recorder to record parameters. Monitorable parameters should be determined by the combination of the two aspects.

Table 3.5 Forms of state output

	No.	State name	Beginning time	Ending time	Relevant parameters
Example	001	Startup	325 s	676 s	Throttle lever, exhaust temperature, mean value of $N1$ speed, etc.
Meaning	The name of the state event numbered 001 is *the state of startup*, which begins at 325 s, the relative time of this flight data record and ends at 676 s. Parameters such as throttle lever, exhaust temperature and mean value of $N1$ speed need special attention				

Table 3.6 Parts of the monitorable parameters for aircraft

No.	Parameters
1	Mean value of high pressure rotor speed at full throttle
2	Mean value of low pressure rotor speed at full throttle
3	Mean value of engine exhaust temperature at full throttle
4	Mean value of high pressure rotor speed at idle
5	Mean value of low pressure rotor speed at idle
6	Mean value of engine exhaust temperature at idle
7	Maximum high pressure rotor speed during takeoff
8	Maximum low pressure rotor speed during takeoff
9	Maximum engine exhaust temperature during takeoff

In a steady state of the aircraft, characteristic parameters are provided by each order statistics information (e.g. mean values, variances and so on), extremums (e.g. maximum value and minimum value) and change information (e.g. changing speed and trend) to describe the features of relevant aircraft subsystems or airborne equipment. Through analysis and induction, parts of the parameters for a type of aircraft which are to be monitored are listed in Table 3.6.

It is observed that parameters which can represent and reflect the performances of major aircraft subsystems can be induced and calculated in a steady state of aircraft. And these parameters can be used as research objects to conduct trend monitoring. This can be beneficial in reflecting the time-domain features of the overall performance of the aircraft and fully taking into account statistics information of flight performance.

3.2.3 Monitoring Method of Mean-Value and Range Based on AR Model

1. Prediction Procedure

$\mathrm{AR}(n)$ model is used to conduct modeling research, which follows time series theory. There are five stages for modeling: data collection, data characteristic test, estimation of model parameters by using estimation methods, model fitness identification and prediction. Reversed order test is used to test data features while modeling, which meets perfectly requirements for stable normal time series with mean value as zero. Least square method is utilized to estimate model parameters and three test criteria of FPE, AIC and BIC are exploited to conduct model fitness test. Results would vary because different orders for search are defined. Therefore, based on the principles of rapid computation, not too high order and practicality, BIC criterion and the search order 60 are employed for the test. Besides, optimum prediction without correction is made to the model. The process of prediction model is shown in Fig. 3.2.

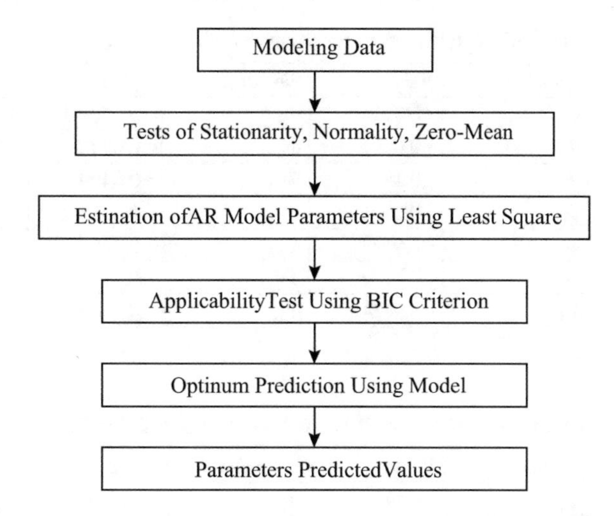

Fig. 3.2 Flight data trend prediction process based on AR model

2. Method for Monitoring

Condition-based maintenance method focuses on monitoring parameter trend, which demands continuous or periodical monitoring of parameters for major functional systems and technical conditions of parts. When the monitored parameters approach the threshold, decisions can be made to change the parts or recover operational capability of the parts. It is a kind of maintenance mode which completes necessary repairs according to actual technical conditions of the objects.

The basic idea of mean-value and range monitoring which focuses on monitoring parameter trend is that when optimum prediction value of the parameters is obtained through $\mathrm{AR}(n)$ model, and this prediction value is fluctuating around the standard value (undoubtedly within thresholds), it indicates that the subsystem which corresponds to the parameter is in a good condition during the next flight when the aircraft is in a steady state and typical flight period. Once the parameter is out of the threshold range, it means malfunction occurs. And if the parameter is close to the threshold, it signifies that preventive measures are required. Besides, the physical meanings of parameters provide scientific basis for rapid positioning of potential failures. Theoretical values and practical maintenance experience are the foundations for determining the standard value and threshold.

3.2.4 Case Study and Effect Assessment

1. Time-Series Modeling

Select 240 sorties of a type of aircraft and take the former 210 data for modeling. Make 30-step forward prediction and compare it with the practical data. Thus prediction error can be calculated. Tables 3.7, 3.8 and 3.9 present orders of the

Table 3.7 AR(15) model for parameter *N*2 at full throttle

Parameter	Value	Parameter	Value	Parameter	Value
Φ_1	0.4748	Φ_8	0.0608	Φ_{15}	−0.0612
Φ_2	−0.0521	Φ_9	0.0740	BIC	−1758.2388
Φ_3	0.1503	Φ_{10}	−0.0499		
Φ_4	0.0091	Φ_{11}	0.0158		
Φ_5	−0.0173	Φ_{12}	0.2256		
Φ_6	0.1102	Φ_{13}	−0.1179		
Φ_7	−0.0124	Φ_{14}	0.1340		

Table 3.8 AR(40) model for parameter *N*1 at full throttle

Parameter	Value	Parameter	Value	Parameter	Value
Φ_1	0.4974	Φ_{16}	0.1711	Φ_{31}	0.1665
Φ_2	−0.1834	Φ_{17}	−0.1878	Φ_{32}	−0.0366
Φ_3	0.1332	Φ_{18}	0.1586	Φ_{33}	−0.0584
Φ_4	0.0682	Φ_{19}	−0.1944	Φ_{34}	−0.1144
Φ_5	−0.0450	Φ_{20}	−0.0739	Φ_{35}	0.0119
Φ_6	0.0651	Φ_{21}	0.12489	Φ_{36}	0.0559
Φ_7	−0.0934	Φ_{22}	−0.1216	Φ_{37}	0.0633
Φ_8	−0.0744	Φ_{23}	0.1092	Φ_{38}	0.1055
Φ_9	0.0895	Φ_{24}	−0.0231	Φ_{39}	−0.0605
Φ_{10}	−0.0702	Φ_{25}	0.0801	Φ_{40}	−0.1047
Φ_{11}	0.0830	Φ_{26}	−0.0664	BIC	−1598.6902
Φ_{12}	−0.0364	Φ_{27}	0.0373		
Φ_{13}	0.0420	Φ_{28}	−0.0349		
Φ_{14}	−0.0738	Φ_{29}	0.0345		
Φ_{15}	−0.0158	Φ_{30}	−0.0480		

Table 3.9 AR(20) model for parameter *T*4 at full throttle

Parameter	Value	Parameter	Value	Parameter	Value
Φ_1	0.5642	Φ_9	0.0513	Φ_{17}	0.0704
Φ_2	−0.1557	Φ_{10}	−0.0083	Φ_{18}	−0.1487
Φ_3	0.1834	Φ_{11}	−0.0387	Φ_{19}	−0.0502
Φ_4	0.0497	Φ_{12}	0.2339	Φ_{20}	−0.0504
Φ_5	−0.0360	Φ_{13}	−0.1078	BIC	−760.2399
Φ_6	0.1788	Φ_{14}	0.2543		
Φ_7	−0.0274	Φ_{15}	−0.0056		
Φ_8	0.0143	Φ_{16}	−0.0728		

parameter models and their values according to BIC applicability criterion, in which $\{a_t\}$ is white noise.

(1) $N2$ at full throttle

$$\text{AR}(15){:}x_t = \phi_1 x_{t-1} + \phi_2 x_{t-2} + \cdots + \phi_{15} x_{t-15} + a_t, a_t \sim \text{NID}(0, 2.711)$$

(2) $N1$ at full throttle

$$\text{AR}(40){:}x_t = \phi_1 x_{t-1} + \phi_2 x_{t-2} + \cdots + \phi_{40} x_{t-40} + a_t, a_t \sim \text{NID}(0, 6.975)$$

(3) T4 at full throttle

$$\text{AR}(20){:}x_t = \phi_1 x_{t-1} + \phi_2 x_{t-2} + \cdots + \phi_{20} x_{t-20} + a_t, a_t \sim \text{NID}(0, 0.0006)$$

A large number of application results of time series models in engineering indicate that the model is available if the prediction error is within 15%. If the error is beyond this value, the prediction is unreliable and new modeling is required. Unequal interval prediction was made to the mean value and range monitoring parameters by using time series models, in which the maximum and minimum of prediction errors were 15 and 0.7%, respectively. This fully validated the built model. Figures 3.3, 3.4, 3.5, 3.6, 3.7, 3.8, 3.9 and 3.10 present the prediction curves of some of the monitored parameters, in which the full lines represent the true values and the dotted lines the prediction values.

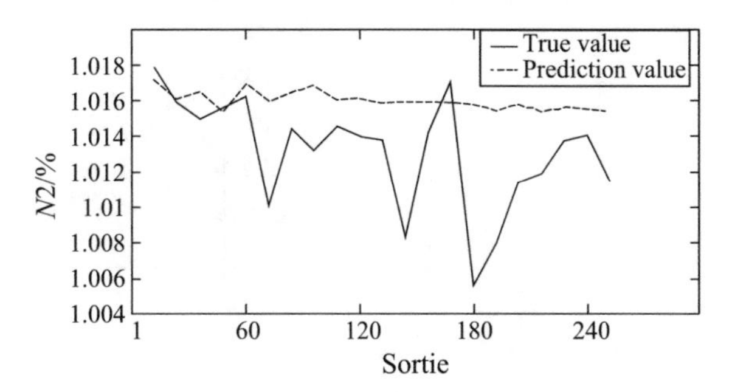

Fig. 3.3 Prediction *curve* of $N2$ at full throttle

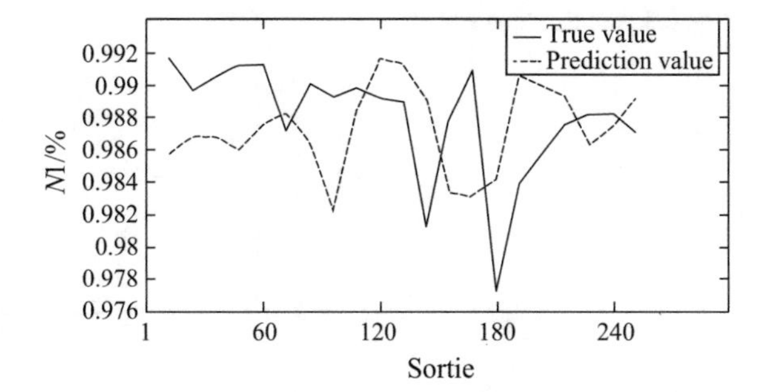

Fig. 3.4 Prediction *curve* of $N1$ at full throttle

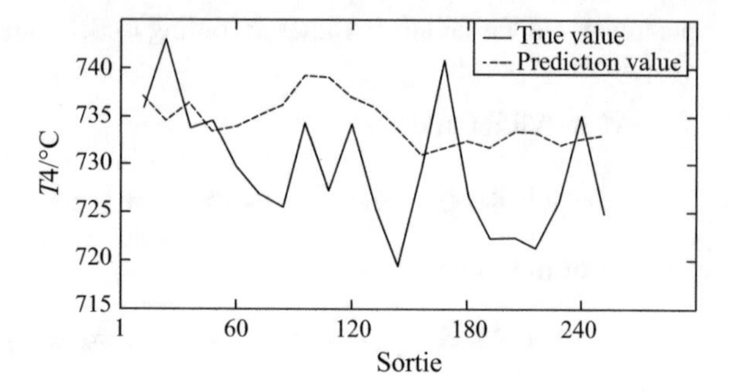

Fig. 3.5 Prediction *curve* of *T*4 at full throttle

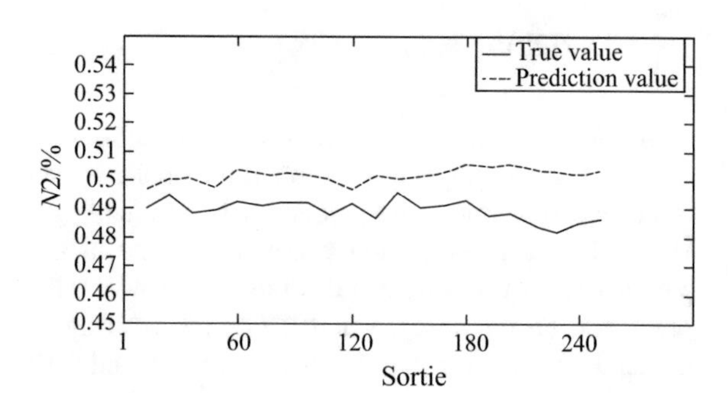

Fig. 3.6 Prediction *curve* of *N*2 at idle

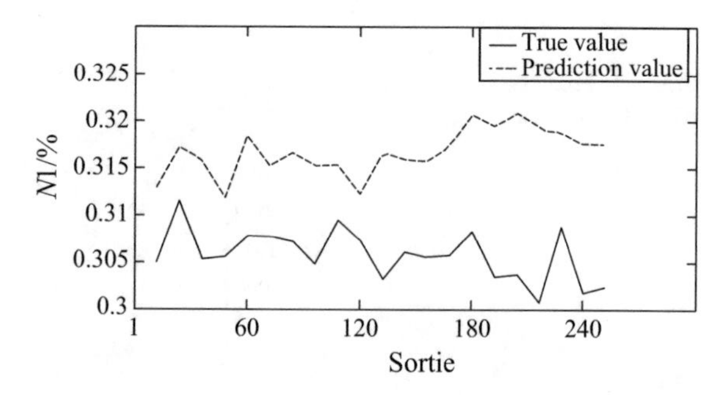

Fig. 3.7 Prediction *curve* of *N*1 at idle

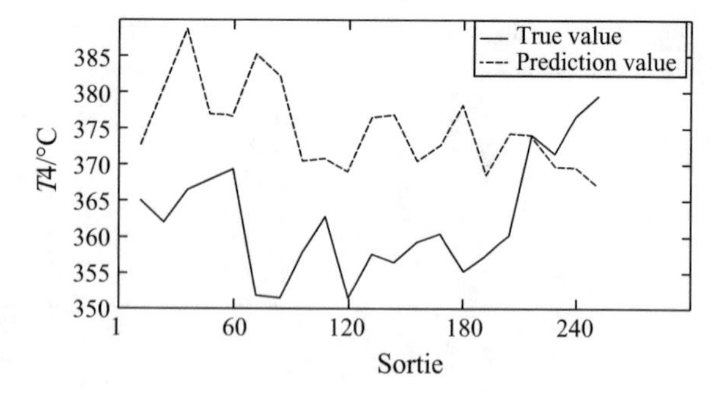

Fig. 3.8 Prediction *curve* of *T*4 at idle

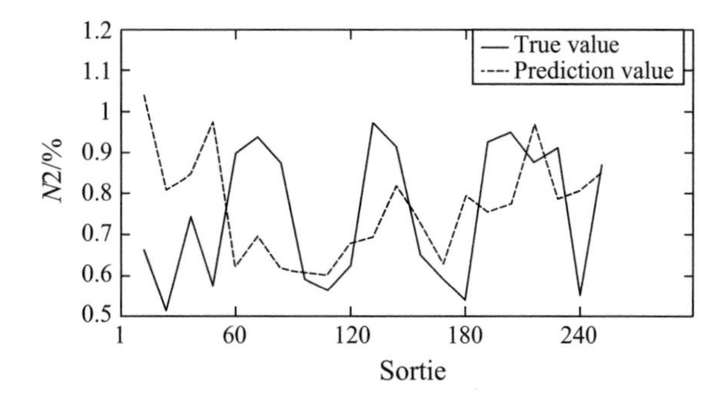

Fig. 3.9 Prediction *curve* of *N*2 at maximum DC voltage

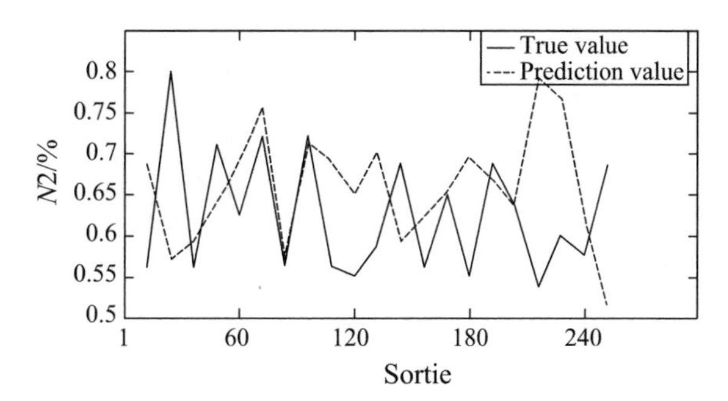

Fig. 3.10 Prediction *curve* of *N*2 at minimum DC voltage

2. Effect Evaluation

Satisfying results were obtained in practical application of this method. Figure 3.11 shows continuous monitoring of *N*1 and *N*2 at full throttle. By analyzing the regular pattern of the curve, we can find that there is a trend for this signal to get close gradually to the standard value from larger errors. Follow-on investigation shows that this is natural for engines because performances of engines are not stable for the opening flights at the beginning of the year, but gradually reach the optimal state after operating for a period of time.

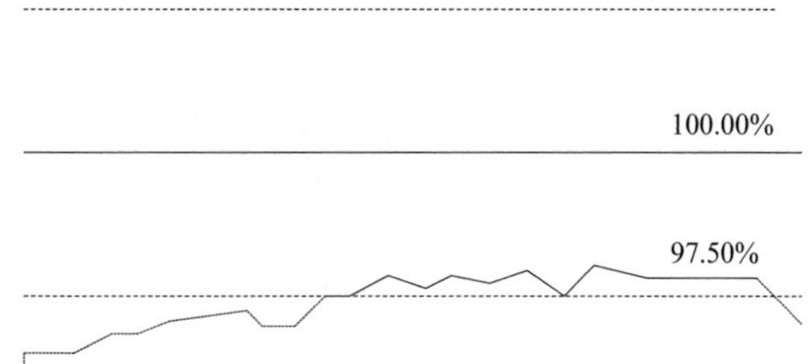

Fig. 3.11 Monitoring *curve* of *N*2 at full throttle

Chapter 4
Similarity Search for Flight Data

Similarity search can be used in search of data necessary for model training, and is therefore, an important part of information mining. Since flight data presents itself as a typical time series, for aircraft and its subsystems in similar operating modes or flight conditions, there should be series sets with similar change trends. By means of searching data series with similar features, similarity search provides technological support for flight data research and "mines out" information of empirical values and their underlying rules, thus laying down a solid basis for flight data reconstruction and monitoring of aircraft conditions. Due to the insufficiency of the traditional method of point-to-point comparison, this chapter will elaborate on unary similarity search of flight data in relation to slope distance, angle distance, and curvature distance. In this chapter, the method of multivariable-series-oriented similarity search will also be introduced. This method, verified by real flight data, makes use of variable step length curve binning and QR decomposition of incidence matrix.

4.1 The Method of Similarity Analysis of Time Series

The method of similarity analysis of time series provides theoretical and methodological basis for time series data mining so that informed decisions can be made in both research and practice. By means of similarity search and series pattern mining, time series can be clustered and classified. By the same token, frequency pattern can be mined out, abnormity pattern identified, and relevance rules extracted.

J. Zhang and P. Zhang, *Time Series Analysis Methods and Applications for Flight Data*, DOI 10.1007/978-3-662-53430-4_4

4.1.1 Overview

Research and application of time series data have been conducted along the following lines: (i) Trend Analysis in which data are analyzed with respect to trend change, cycling characteristics, seasonal change, and random change as well; (ii) Similarity Search which deals with algorithm design for similarity search and language design for similarity query; (iii) Sequence Pattern Mining which, by means of data mining, identifies patterns occurring more frequently than others, or than at other time points; (iv) Periodic Analysis concerns the mining out of periodic patterns, that is, sorting out from time series data the recurrent patterns.

Similarity, search of time series, also referred as similarity query, means the search for the data series that is closest to the given sequence. Since 1993 when Agrwal et al. published the first paper on this topic, similarity search of time series database has attracted both attention and effort from scholars in the field of database and data mining. The research focus is on dimension reduction method, similarity measurement, and similarity match and search. As a result, the theory and practice of similarity analysis of time series has grown into an important research area of data mining and intelligent data processing. In 2003, Robert F. Engle and Clive W. J. Granger, economists from America and Britain respectively, shared the Nobel Prize between them for their contribution in the study of time series in analytic economy. In China and supported by 863 National Program, two research programs, Shape Representation-Based Similarity Search of Time Series and Pattern Distance of Time series, were accomplished successively in 2000 and 2004, paving the way for further application and research of similarity of time series.

Similarity search of time series can be divided into two types: full series match, in which the series to be searched and the series recorded in database fully match in length; and subseries match in which series to be searched is shorter than the recorded series. Since "slide window" technology can be used to transform subseries match into full series match, only the latter is discussed in the present study.

The study of full series match covers three research areas:

(1) All-Pairs Query of similarity pattern in time series. In a time series database $x_1, x_2, \ldots, x_n$, when similarity measurement function D and threshold ε are given, identify all series pairs with mutual distance smaller than ε ($\varepsilon > 0$).

(2) Range Query of similarity pattern in time series. In a time series database $x_1, x_2, \ldots, x_n$, and for standard query series $\boldsymbol{Q}$, when similarity measurement function $\boldsymbol{D}$ and threshold ε are given, identify all the series x_i whose distance to $\boldsymbol{Q}$ is smaller than ε: $D(Q, x_i) \leq \varepsilon$.

(3) k-Nearest Neighbor Query of similarity pattern in time series. In a time series database $x_1, x_2, \ldots, x_n$, for standard query series $\boldsymbol{Q}$ and the given similarity measurement function $\boldsymbol{D}$, identify k series x_i that are in the smallest distance to $\boldsymbol{Q}$: $\max(D(Q, x_i)) = \varepsilon$.

Similarity search aims at solving two basic problems: establishment of index structure and completeness of query, both of which rely on effective time series dimension reduction and similarity measuring.

4.1.2 Time Series Dimension Reduction

For similarity search to be done, comparison should be made between the preset series for query and all the series in the database. If the given series length is $\boldsymbol{N}$, and there are n databases, the computation complexity will be $\boldsymbol{O}(\boldsymbol{N_n})$, or even $\boldsymbol{O}(\boldsymbol{N_n^2})$. The increasing series length may lead to "dimension explosion," which is not acceptable in practice. Therefore, one of the key issues in solving problems in similarity search is a method characterized with strong robustness and low computation complexity.

Dimension reduction of time series begins with pattern representation. The logic underlying pattern representation goes like this: extracting features of times series, transforming them into the feature space, and expressing original time series in terms of feature patterns of the feature space. This way of expression has two advantages: compression of time series data helps to make effective data mining; preservation of major features and deletion of trivial details better reveals the changing of time series, and therefore, improves data mining.

1. Discrete Fourier Transform (DFT)

DFT is the earliest method used to represent dimension reduction of time series. As an algorithm, DFT is a transformation independent of data. If the time domain distance between two signals is equal to their Euclidean distance in the frequency domain, and at the same time, lower bound theorem is satisfied, there will not be false-negative search results. Since the first several coefficients of DFT carry most of the signal energy, dimension reduction can be realized by means of preserving these coefficients, while using Euclidean distance as similarity measurement of time series. If the computed distance is smaller than the threshold value preset by the user, it can be concluded that the two series are similar to each other. Moreover, $\boldsymbol{O}(N\text{lg}N)$, time complexity of DFT, proves to be better than $\boldsymbol{O}(\boldsymbol{MN})$ or even $\boldsymbol{O}(\boldsymbol{MN}^2)$, which is the time complexity of a point-to-point comparison. In other words, DFT costs much less computation time.

The DFT algorithm goes like this: when signal $\vec{x} = [x_t], t = 0,\ldots,n-1$, DFT at point n can be defined as series $\boldsymbol{X}_f$, which consists of n complex numbers $(f = 0,\ldots,n-1)$:

$$\boldsymbol{X}_f = \frac{1}{\sqrt{n}}\sum_{t=0}^{n-1} x_t \exp\left(-\frac{j2\pi ft}{n}\right) \quad f = 0,\ldots,n-1, \tag{4.1}$$

where j is an imaginary unit and $j = \sqrt{-1}$. Meanwhile and as shown below, signal x can be recovered through inverse transformation:

$$x_t = \frac{1}{\sqrt{n}} \sum_{f=0}^{n-1} X_f \exp\left(\frac{j2\pi ft}{n}\right) \quad t = 0, \ldots, n-1$$

The original signals in Fig. 4.1 cover 4096 data points. After applying DFT to these data points, graphical recovery is done, respectively, to the first 32 and 64 transformation coefficients. The recovery results are shown in (a) and (b) of Fig. 4.2.

A comparison between Figs. 4.1 and 4.2 shows that when more DFT coefficients are preserved, more local features will be preserved; and that DFT may cause information loss as a result of omission of high-frequency elements of signals and smoothing of local extremums of times series. And meanwhile, since DFT requires a high stability of time series, there should be limitations in its actual use. Even though some later researchers have proposed DFT-based extended or improved algorithms, the central idea remains the same.

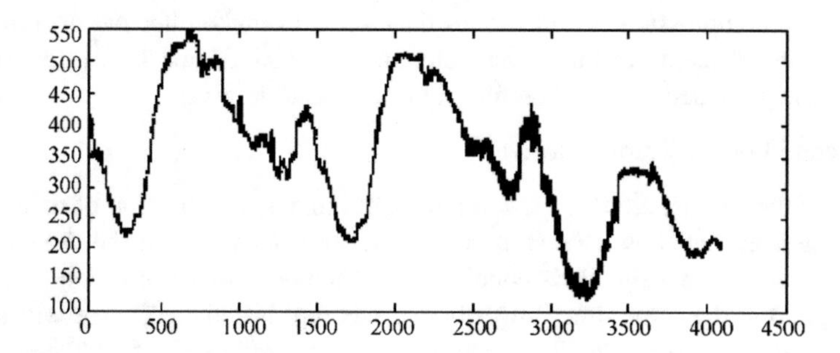

Fig. 4.1 Original signals

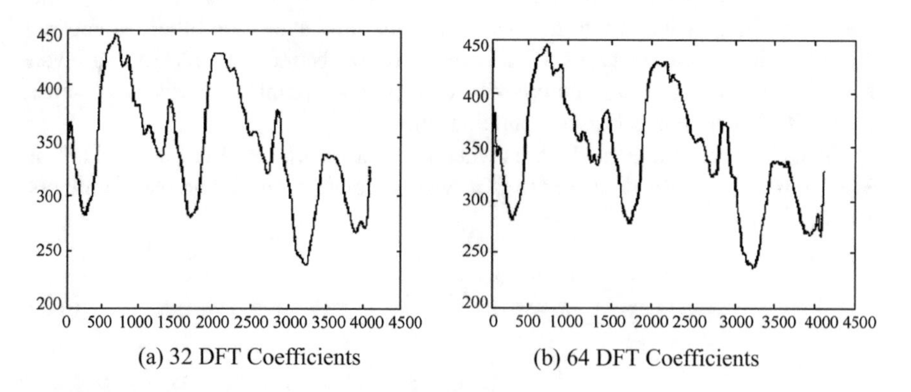

Fig. 4.2 Graphical representations of DFT coefficients

2. Discrete Wavelet Transform (DWT)

In addition to its good locality, DWT is also characterized with simultaneous reflection of signal differences in both time and frequency domain. The application of DWT in similarity search of time series was first proposed by Chan et al. in 1999, which was followed by some research articles on DWT-based similarity search and measurement.

Similar to DFT, DWT realizes dimension reduction by only preserving strong wavelet coefficients, but it is a relatively better lossy compression. If the same number of coefficients is preserved by DWT and DFT, DWT provides more precise approximation than the original data do. More importantly, owing to its good locality, DWT can preserve local details simultaneously in time and frequency domain. Moreover, as an algorithm with $O(N)$ as its time complexity, DWT works better than the point-by-point search and DFT.

As far as data processing is concerned, DWT is highly similar to DFT, its basic function being defined by the recursive function as follows:

$$\psi_{j,k}(t) = 2^{\frac{j}{2}}\psi(2^j t - k) \tag{4.2}$$

In this expression, 2^j is the scale of t (while j is the logarithm of the scale with 2 as the base); $2^{-j}t$ indicates transformation in time domain while $2^{\frac{j}{2}}$ keeps the wavelet standard of L^2 (space of square integrable function) at different scaling ratios. Thus, any function within $\boldsymbol{L}^2(\boldsymbol{R})$ space can be expressed as a series as shown below:

$$f(t) = 2\sum_{j,k} a_{j,k} 2^{\frac{j}{2}}\psi(2^j t - k) \tag{4.3}$$

In combination with (4.2), (4.3) is converted into:

$$f(t) = 2\sum_{j,k} a_{j,k}\psi(t) \tag{4.4}$$

In (4.4), the series made up of the two-dimensional coefficients is referred to as DWT of $\int(t)$. The algorithm of $a_{j,k}$ can be expressed through inner product as follows:

$$f(t) = \sum_{j,k} \langle \psi_{j,k}(t), f(t) \rangle \psi_{j,k}(t)$$

As for the original signals shown in Fig. 4.1, applying Harr DWT to them and preserving, respectively, 32 and 64 DWT coefficients, the graphical representation of which is shown in Fig. 4.3 as follows.

Compared with the curve of DFT dimension reduction, DWT curve is not as smooth. The main reason for this DFT curve is obtained through capturing the first

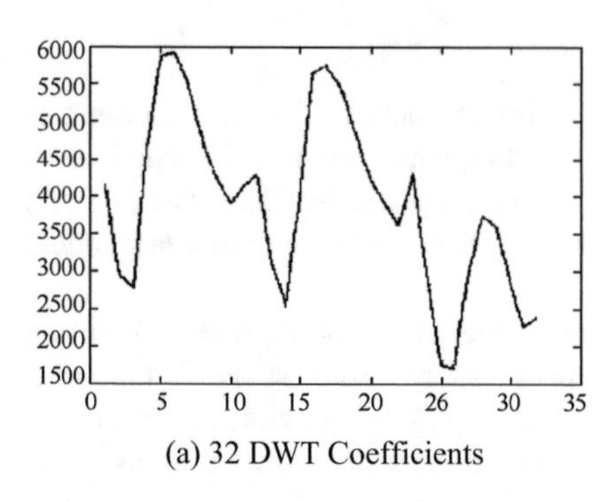

(a) 32 DWT Coefficients

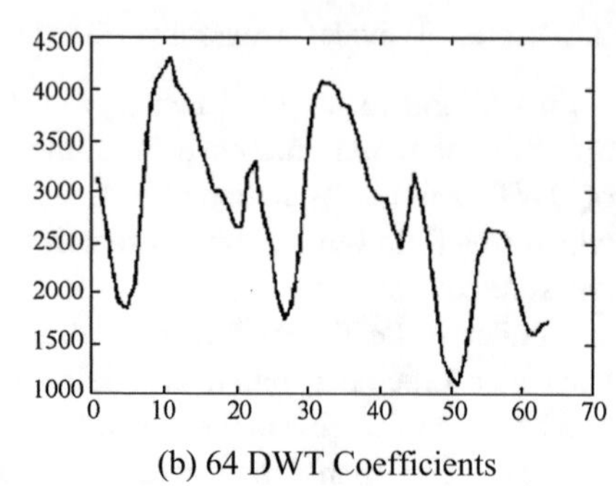

(b) 64 DWT Coefficients

Fig. 4.3 Graphical representation of Harr DWT

few transformation coefficients, to which inverse transformation is then applied. In other words, the data points making up the curve remain the same in quantity, that is, 4096. For DWT, however, since in each transformation the number of the preserved the global data is half of that before transformation, the number of the elements used for curve description is 16, 32, and 64. It is for this reason that DWT curve shows sharper changes. In fact, DWT does better in maintaining local features of the data under processing.

Previous researches have shown that in similarity search of time series DFT and DWT do not make much difference. According to Li-Leh Wu, DWT neither reduces relative mirror error, nor does it improve precision of similarity query. Additionally, due to DWT's incapability of processing series of random length, its actual use is by no means flawless. Therefore, DFT in combination with DWT is expected to work better than separate use of the two methods. Keogh, for example, in his study of similarity search of time series, proposed a method which combines DFT and DWT and takes advantages of both the methods. It is shown that Keogh's solution not only produces better results but also takes less time.

3. Singular Value Decompression (SVD)

SVD is a common method of dimension reduction and has been widely in image and text indexing. Also known as Principal Component Analysis (PCA) or Kullback–Leibler decomposition in statistics, SVD is a projection method based on distribution of statistical probability, and is used in the analysis of noisy and height-related measured data. SVD projects high-dimensional information onto low-dimensional subspace, and preserves major process information. In other words, in an SVD process, first, N n-dimensional orthogonal vectors that best represent the data are identified ($N \leq n$), and then, the original data is projected into a smaller space, and consequently compressed or dimension-reduced.

Suppose X represents a data matrix $n \times m$ in which the row corresponds the samples, and the column to the variables. The matrix can be decomposed into the following:

$$X = TP^{\mathrm{T}} = t_1 p_1^{\mathrm{T}} + t_2 p_2^{\mathrm{T}} + \cdots + t_m p_m^{\mathrm{T}} \tag{4.5}$$

where $T = [t_1, t_2, \ldots, t_n]$ is referred to as score matrix; $t_i \in R^n$, $i = 1, 2, \ldots, n$ as score vector, or principal component of X; $P = [p_1, p_2, \ldots, p_m]$ as load matrix; and $p_i \in R^m$ as load vector ($i = 1, 2, \ldots, m$).

The score vectors are in an orthogonal relationship and so are the load vectors. If both sides of Eq. (4.5) right multiply P_i, it will be converted into:

$$Xp_i = t_1 p_1^{\mathrm{T}} p_i + t_2 p_2^{\mathrm{T}} p_i + \cdots + t_m p_m^{\mathrm{T}} p_i = t_i$$

It can be seen that the projection of data matrix X in the direction of load vector P_i is actually the corresponding score vector t_i, the length of which ($\|t_i\|$) indicates the coverage of the matrix in the direction of P_i. If a degree of linear dependence exists between matrix variables, the matrix variation will show itself mainly in the direction of the first k load vectors, and there will be very little projection for the final load vector (owing to measuring noise and system errors). Thus, the data matrix can be decomposed into:

$$X = t_1 p_1^{\mathrm{T}} + t_2 p_2^{\mathrm{T}} + \cdots + t_k p_k^{\mathrm{T}} + E = T^{\mathrm{T}} P + \widetilde{T}\widetilde{P}^{\mathrm{T}} \tag{4.6}$$

In this expression, $P = [p_1, p_2, \ldots, p_k] \in R^{m \times k}$ is the principal load matrix; $\widetilde{P} = [p_{k+1}, p_{k+2}, \ldots, p_n] \in R^{m \times (n-k)}$, residual load matrix; $T = [t_1, t_2, \ldots, t_k] \in R^{n \times k}$, principal score matrix; $t_1, t_2, \ldots, t_k$ and $\|t_1\| \geq \|t_2\| \geq \cdots \geq \|t_k\|$, principal load vector; and $\widetilde{T} = [t_{k+1}, t_{k+2}, \ldots, t_n] \in R^{n \times (m-k)}$, residual score matrix.

Also in this equation, $[P, \widetilde{P}]$ is a unit orthogonal matrix, $[T, \widetilde{T}]$ an orthogonal matrix, while TP^{T} is the principle component space, and the projection of X in the direction of principal load vector. $E = \widetilde{T}\widetilde{P}^{\mathrm{T}}$ is the residual space, which is the projection of X in the direction of all the load vectors other than principal components. Since error matrix E is mainly a result of measurement noise and system error, noise interference can be eliminated by canceling E, and elimination will not cause any loss of effective information. Consequently, matrix X can be approximately reexpressed as follows:

$$X \approx t_1 p_1^{\mathrm{T}} + t_2 p_2^{\mathrm{T}} + \cdots + t_k p_k^{\mathrm{T}} = TP^{\mathrm{T}} \tag{4.7}$$

It can be proved that the load vector of matrix X is actually the eigenvector λ_i of covariance matrix $X^{\mathrm{T}}X$, and that the principal component analysis of X is actually the analysis of λ_i.

The SVD expression of time series is different from DFT and DWT, since it is an overall reexpression of a time series database, and extracts and transforms features of the database as a whole. For SVD, this type of overall transformation is where its strength and weakness lie. On the one hand, SVD as a linear transformation minimizes error in data reconstruction and works better than other transformations in many instances; on the other hand, $\boldsymbol{O}(\boldsymbol{M}n^2)$, SVD time complexity, means a higher computation cost because anytime when a time series is inserted or deleted, SVD expression of the whole time series database must be recomputed. In $\boldsymbol{O}(\boldsymbol{M}n^2)$, $\boldsymbol{M}$ is the size of the database while n the mean length of time series. A great number of SVD algorithms have been proposed, but they all share the problem of higher computation cost. What makes things worse is that some fast algorithms have "reluctantly" introduced into themselves the probability of abandoning "unacceptable errors."

4. Piecewise Linear Representation (PLR)

PLR, a widely used representation method, makes use of a set of head-to-tail adjacent line segments to approximately represent time series, and has some characteristics and advantages unavailable in other methods. As early as in 1974, Pavlidis and Horowitz pointed out that PLR is an effective method of data compression and noise elimination, featuring multi-granularity analysis in time domain. In PLR of time series, the approximate granularity to the original time series is determined by the number of line segments. The more line segments there are, the smaller their average length will be; so a larger segment number reflects more about short-term fluctuation of time series. The fewer line segments there are, the longer their average length will be; so a smaller segment number reveals more of medium and long-term fluctuation of time series.

Following the introduction, by Keogh et al., of such segmentation methods as Piecewise Aggregate Approximation (PAA) and Adaptive Piecewise Constant Approximation (APCA), PLR has become one of the key models of time series representation. According to these models mentioned above, first, a time series is divided into several segments, which are then averaged. As for the representation method, weighting Euclidean distance is used, which improves speed and accuracy of the method. Additionally, time series of arbitrary length can be processed and continuous insertion and deletion of time series is allowed.

Suppose $\boldsymbol{X} = x_1, \ldots, x_n$ is a time series, and $\boldsymbol{Y} = \{\boldsymbol{Y}_1, \ldots, \boldsymbol{Y}_k\}$ a time series database. Without loss of generality, let the length of every series in $\boldsymbol{Y}$ be n, and $\boldsymbol{N}$ be the number of dimensions of the transformation space, and $1 \leq N \leq n$. And then suppose $\boldsymbol{N}$ is a submultiple of n, the supposition not being the condition of the method, but for simplification of symbols.

Time series with length n can be expressed through $\boldsymbol{N}$-dimension space vector: $\overline{\boldsymbol{X}} = \overline{x_1}, \ldots, \overline{x_N}$, and the ith element of $\overline{\boldsymbol{X}}$ can be calculated through the following equation:

$$\overline{x_i} = \frac{N}{n} \sum_{j=\frac{n}{N}(i-1)+1}^{\frac{n}{N}i} x_j$$

Figure 4.4 is an illustration of the segmented data reduction. To put it in a simple way, for data to be reduced from n-dimension vectors to N-dimension ones, first, it is segmented into N segments of equal length, and then the mean value of each

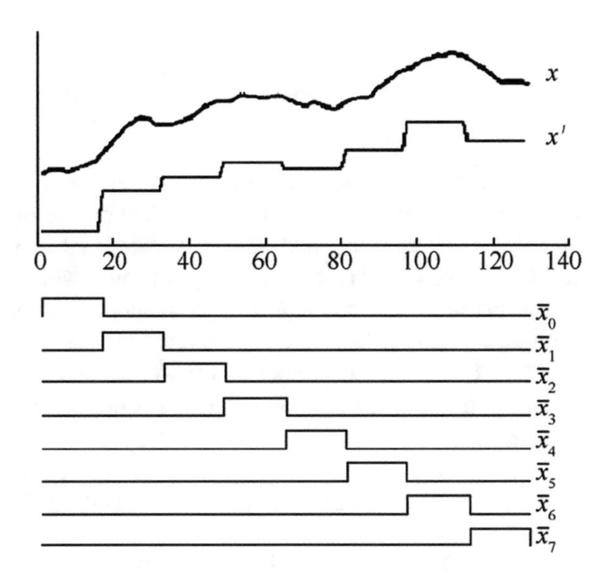

Fig. 4.4 Illustration of PAA dimension reduction

segment is calculated. These mean values make up the vector, which is actually the representation of the original time series after dimension reduction. When $N = n$, the transformed representation is the same as the original one; when $N = 1$, the transformed result is the simple arithmetic mean of the original time series. Generally, what is obtained through this type of transformation is the piecewise constant approximation of the original series. This presentation method is therefore termed as PPA by Keogh.

Figure 4.5 shows the result of application of PAA to the signals in Fig. 4.1, with the first 64 data preserved. As indicated by Fig. 4.5, PAA smoothes the local features of time series data as DFT does: the higher the change frequency of the original data, the larger the change amplitude, and consequently the better smoothing effect. But at the same time, there will be more information loss and errors.

As for similarity measurement in PAA, in order not for error abandonment to occur, Keogh proposed a distance measurement (DR) which is defined in the index space and satisfies the lower boundary theorem $\text{DR}(\bar{x}, \bar{y}) \leq D(X, Y)$. This distance measurement can be expressed as follows:

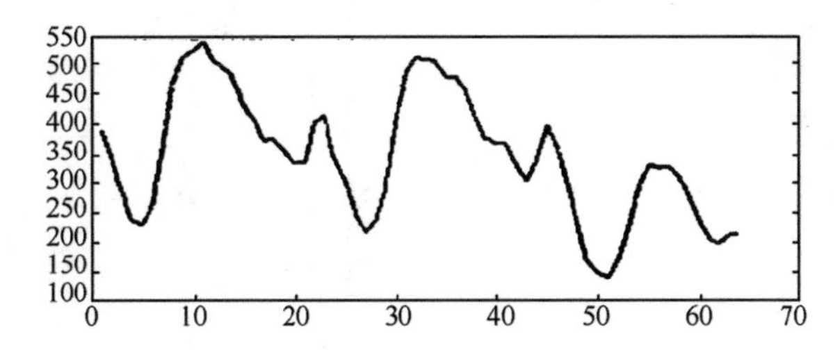

Fig. 4.5 Recovery graph of 64 PAA segments

$$\mathrm{DR}(\overline{x},\overline{y}) = \sqrt{\frac{n}{N}}\sqrt{\sum_{i=1}^{N}(\overline{x_i}-\overline{y_i})^2}$$

Since PAA is free of error abandonment and features high precision and flexible processing objects, it may after all be a good processing method. However, PAA is also a coarse-grained method, since the same length is extracted from each segment for the sake of a convenient indexing in its use, which will very probably give rise to misreporting and subsequent computational complexity. For example and as shown in Fig. 4.6, there are two given series $\boldsymbol{X}$ = (−1, −2, −1, 0, 1, 1, 0) and $\boldsymbol{Y}$ = (2, −1, −2, −3, −1, 1, 2, 2), and PAA of $\boldsymbol{X}$ and $\boldsymbol{Y}$ are, respectively, expressed as $\overline{\boldsymbol{X}}$ and $\overline{\boldsymbol{Y}}$. If both series are segmented into two pieces, there will be the following equations:

$$\overline{\boldsymbol{X}} = (\mathrm{mean}(-1,-2,-1,0),\mathrm{mean}(2,1,1,0)) = (-1,1)$$

$$\overline{\boldsymbol{Y}} = (\mathrm{mean}(2,-1,-2,-3),\mathrm{mean}(-1,1,2,2)) = (-1,1)$$

Since $D(\overline{X},\overline{Y}) = 0$, $\boldsymbol{X}$ and $\boldsymbol{Y}$ enter into similar candidate sets, and therefore, misreporting may come as a result.

To overcome the above-mentioned PAA weak points, Keogh later proposed an improved method which can deal with the processing of pieces of arbitrary length. This method is what Keogh referred as Adaptive Piecewise Constant Approximation (APCA). For every segmented piece of a time series, APCA records two values, i.e., the mean value of the all the data points, and the length. Take a given time series $\boldsymbol{c} = \{c_1,\ldots,c_n\}$ for example, its APCA transformation can be expressed as follows:

$$\boldsymbol{C} = \{\langle \boldsymbol{cv}_1,\boldsymbol{cr}_1\rangle,\ldots,\langle \boldsymbol{cv}_M,\boldsymbol{cr}_M\rangle\},\boldsymbol{cv}_0 = 0$$

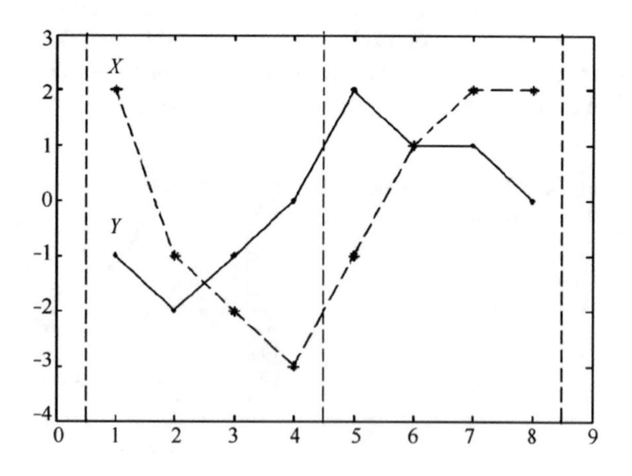

Fig. 4.6 Weak points of PAA transformation

In this expression and for the *i*th segment of the time series, $\boldsymbol{cv}_i$ is the mean value of all the data points, while $\boldsymbol{cr}_i$ the endpoint of the piece.

Through APCA, time series $\boldsymbol{c}$ is transformed into time series $\boldsymbol{C}$ accompanied by an query series $\boldsymbol{Q}$. It is evident that since $\boldsymbol{C}$ carries less information than $\boldsymbol{c}$ does, it is impossible to precisely define between $\boldsymbol{Q}$ and $\boldsymbol{C}$ a distance measurement equivalent to the Euclidean distance. Thus, Keogh and others have defined two types of distance measurement between $\boldsymbol{Q}$ and $\boldsymbol{C}$ as an approximation of $\boldsymbol{D}(\boldsymbol{Q}, \boldsymbol{C})$: one is $\boldsymbol{D}_{\text{AE}}(\boldsymbol{Q}, \boldsymbol{C})$, which is in strict accordance with Euclidean distance; the other is $\boldsymbol{D}_{\text{LB}}(\boldsymbol{Q}, \boldsymbol{C})$, which is loose accordance with Euclidean distance. In specific terms, $\boldsymbol{D}_{\text{AE}}(\boldsymbol{Q}, \boldsymbol{C})$ and $\boldsymbol{D}_{\text{LB}}(\boldsymbol{Q}, \boldsymbol{C})$ can be defined by the equations given below:

$$D_{\text{AE}}(\boldsymbol{Q}, \boldsymbol{C}) = \sqrt{\sum_{i=1}^{M} \sum_{k=1}^{\sigma_i - \sigma_{i-1}} \left(\boldsymbol{cv}_i - \boldsymbol{q}_{k+cv_{i-1}}\right)^2}$$

$$D_{\text{LB}}(\boldsymbol{Q}', \boldsymbol{C}) = \sqrt{\sum_{i=1}^{M} (\boldsymbol{cr}_i - \boldsymbol{cr}_{i-1})^2 (\boldsymbol{qv}_i - \boldsymbol{cv}_i)^2}$$

In these two expressions,

$$\boldsymbol{Q}' = \{\langle \boldsymbol{qv}_1, \boldsymbol{qr}_1 \rangle, \ldots, \langle \boldsymbol{qv}_M, \boldsymbol{qr}_M \rangle\}, \quad \boldsymbol{qv}_i = \boldsymbol{cv}_i, \quad \text{and} \quad \boldsymbol{qv}_i = \frac{\sum_{i=cv_{i-1}+1}^{cv_i} \boldsymbol{q}_i}{\boldsymbol{cr}_i - \boldsymbol{cr}_{i-1}}$$

As shown in Figs. 4.7 and 4.8 illustrate the comparison between the Euclidean distance and the two APCA distances.

Another piecewise linear representation is referred to as Piecewise Polynomial Representation (PPR), which is based on an orthogonal transformation of

Fig. 4.7 D_{AE} and D_{LB}

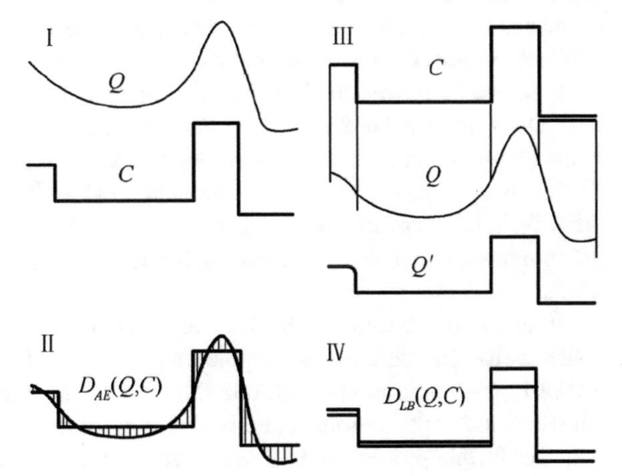

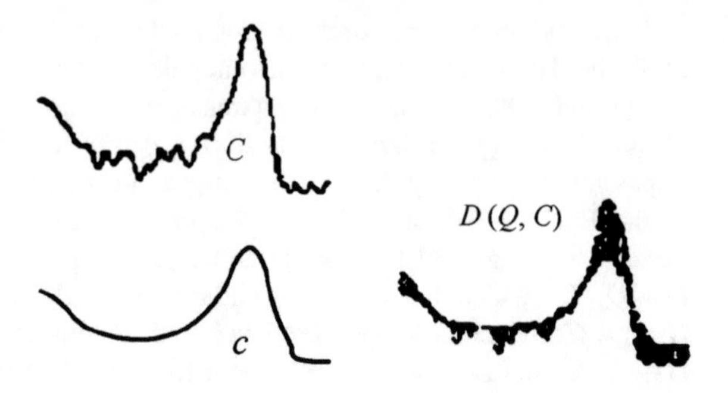

Fig. 4.8 Euclidean distance

linear- polynomial regression. PPR makes use of polynomial function rather than segmented pieces for the representation of time series during a certain period of time. When the polynomial order is 1, PPR regresses to piecewise representation of time series.

5. Landmark Mode

Proposed by Perng et al., Landmark Mode (Landmark for short) is a representation method which integrates within itself similarity model and data model. This method seems to be inspired by the human cognitive experience that two curves are considered similar if they extend and zigzag in a similar way. The central idea of Landmark is that some very important points of the lines can be defined as boundary marks which can be used in place of the original data to be processed. It is in this way that data compression is realized. In different applications of Landmark, different points are selected as landmarks. They may be as simple as the local maximum or minimum, or an inflection point; they may also be as complicated as a complex pattern. Take a curve for example, if the point at which the *n*th derivative is 0 is called the *n*th landmark of the curve, the point where the local maximum or minimum is will be the 1st landmark of the curve, and the inflection point will be the 2nd landmark. But in comparison of similarity, high landmarks have very little influence on the change of time series, and therefore, data compression can be realized by abandoning high landmarks.

It can be seen from the introduction above that Landmark is different from DFT and DWT in that DFT and DWT smooth the of local maximum and minimum points while Landmark preserves most the local features of the original signals. As is shown by Fig. 4.9, after Landmark is applied to the signals in Fig. 4.1, 2335 out of 4096 data points are preserved, compression rate being lower than 50%. In terms of compression efficiency, Landmark Mode is not as effective as DFT, or DWT, or PPA.

In order to solve the problem of low compression efficiency, Landmark eliminates noise interference by employing Minimal Distance/Percentage Principle (MDPP) as a smoothing method. Compared with the other methods mentioned above, Landmark is problematic in one way or another. On one hand, Landmark requires further process of data, and therefore increases computation complexity; on the other hand, since it is hard to predict the final length of the compressed data,

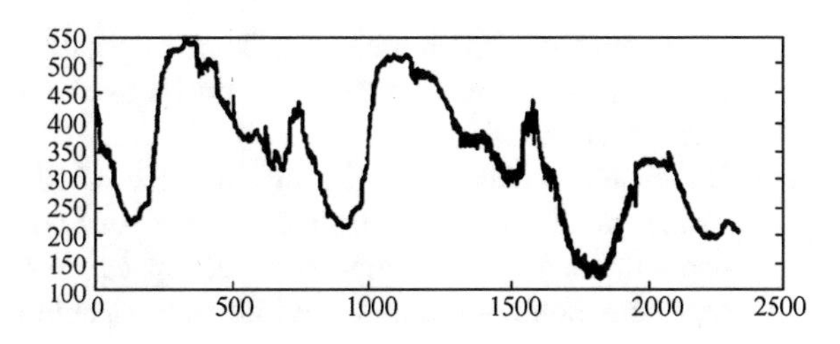

Fig. 4.9 Extraction result of landmark

Landmark does not do very well in satisfying the requirements for similarity measurement.

6. Symbolic Representation

Symbolic representation projects, in a discrete manner, the real-value of time series or its wave form of a certain time period onto a table of limited symbols. In this way, a time series is re-repressed as an ordered set of limited symbols, i.e., a character string. Although this method takes advantage of the recent achievements made in study of character string, it is difficult to decide a suitable discrete algorithm by means of which symbols are interpreted, and similarity measurement is defined.

7. Selection of Dimension Reduction Method

The representation methods introduced so far have revealed the general consensus in the study of similarity search of time series, that is, the similarity means much more than the inter-relationship between two time series. From the initial point-to-point comparison to DFT and DWT transformation, and later on to various methods of piecewise representation or Landmark, or even far later to combinations of multiple methods, the choice of time series representation has become a considerably subjective matter. It may be a subjectively preferred linear transformation, DFT or DWT for example. It may also be personally favored features or key points, which are processed in place of the original series, for example, piecewise processing, key features, slope, and landmarks as well. The chosen features or points will work in cooperation with a suitable similarity measurement function.

Although the algorithm of time series representation is to some extent a subjective choice, this does not mean that time series accept any type of representation. A qualified method of representation should have the following qualities:

(1) Precision. Whichever type of representation or transformation is chosen, it should be able to reduce to the least information loss. It should also be able to account for the change trend of the time series, and at the same time, account for the local features of the signals as precisely as possible.
(2) Rapidity. In similarity search of a large size time-series database, a representation algorithm with good time complexity is often expected for the sake rapidity.
(3) Uniformity. Since the major purpose of an algorithm of time series representation is the comparison for similarity, there should be consistent similarity

measurement for both transformed and original time series, in other words, the same similarity measurement should be used for the time series before and after transformation.

(4) Dimension reduction of the original series. High dimension is often what makes similarity search in time series a very complicated matter. In view of computational time complexity and space complexity of storage, a good representation algorithm should be well capable of dimension reducing.

In making the choice of a representation algorithm, efforts should be made to coordinate computational complexity, effectiveness of dimension reduction, as well as extraction of local and overall features so as to improve the efficiency of dimension reduction. One possible way to evaluate a representation algorithm is to see whether it is in waveform with the human cognitive mechanism of pattern recognition. When people observe a graphic pattern, in most cases, what comes into their view first is "what it is," rather than "how it is," in other words, their attention goes first to the general appearance and then to the details. Time-series data can be seen as a waveform, which is observed in a similar way as is a graphic pattern. Accordingly, for an effective observation of this wave form, it is recognized in the order of importance of the wave parts, that is, first some important points that decide the wave shape, and then the local details. This is an "up-down" method which identifies and analyzes a wave form by means of piecewise linearization. Since this method follows the general rule of human cognition, it works faster than others. So it is advantageous to work out an representation algorithm which is based on the mechanism of human cognition.

4.1.3 Similarity Measurement of Time Series

Second to dimension reduction, similarity measurement is another basic issue in data mining of time series. Since no two pieces of time series are completely identical, the degree of identity must be measured in terms of a value of similarity. Due to the high complexity of time series, the measurement must be in maximum support deformations like amplitude shifting and scaling, linear drift, and time scaling. Similarity measurement must meet the following requirements:

(1) Similarity measurement should allow inaccurate mach and support various types of deformation of time series;
(2) Similarity measurement should have high computational efficiency;
(3) Similarity measurement should enable quick indexing.

For a detailed introduction of similarity measurement, some concepts are worthy of notice since they are often used in time series analysis.

(1) Amplitude shifting. As shown in Fig. 4.10, amplitude shifting occurs when two time series of similar pattern fluctuate at different horizontal base lines.

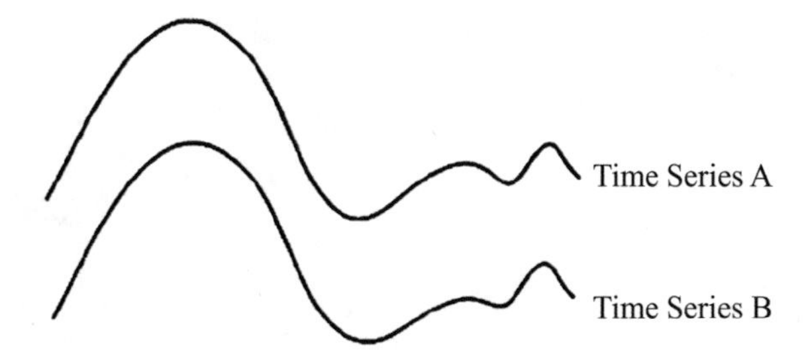

Fig. 4.10 Amplitude shifting of time series

There is often a big ordinary norm distance between two time series with different horizontal base lines, which is contradictory to their similarity in pattern.

(2) Amplitude Scaling. As shown in Fig. 4.11, amplitude scaling occurs when two time series fluctuate at different amplitudes. Since there is also a big ordinary norm distance between them, the similarity can hardly be precisely described.

(3) Discontinuity. As show in Fig. 4.12, discontinuity occurs when two time series are similar in wave form except for interruptions at very few time points or slots. Since interruption of this type is highly sensitive to ordinary norm distance, it will cause a sharp increase of the latter, and consequently similarity between the two time series will not show itself as it actually is.

(4) Linear Drift. As shown in Fig. 4.13, linear drift occurs when two time series are similar, but because of the effect of certain linear factors, the value of one of them shows a trend of progressive increase or decrease. The mathematic expression of linear drift can be approximately presented as follows:

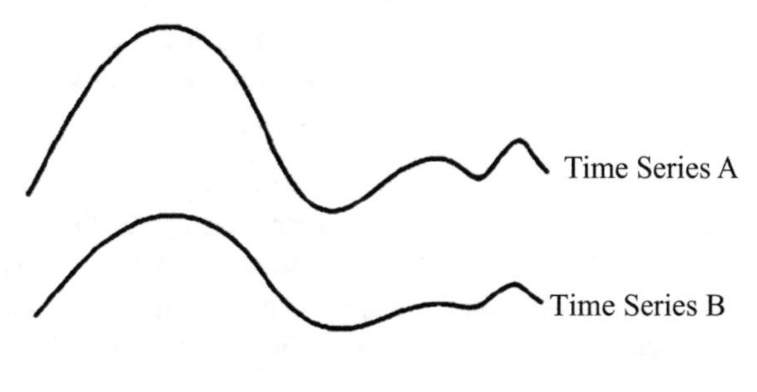

Fig. 4.11 Amplitude scaling of time series

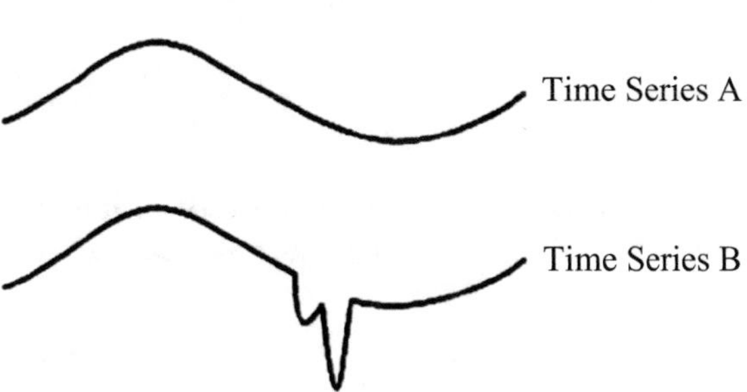

Fig. 4.12 Discontinuity of time series

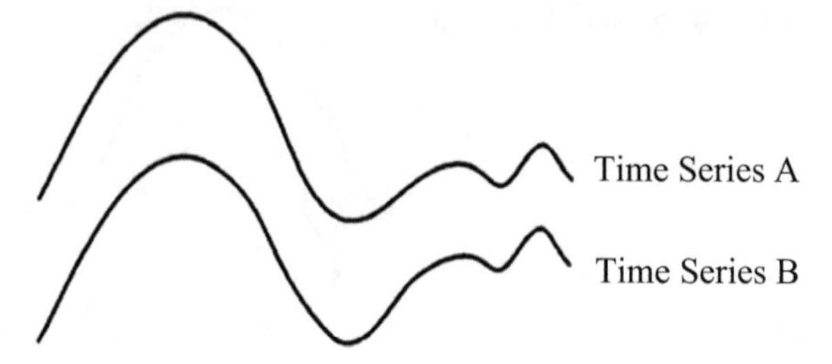

Fig. 4.13 Linear drift of time series

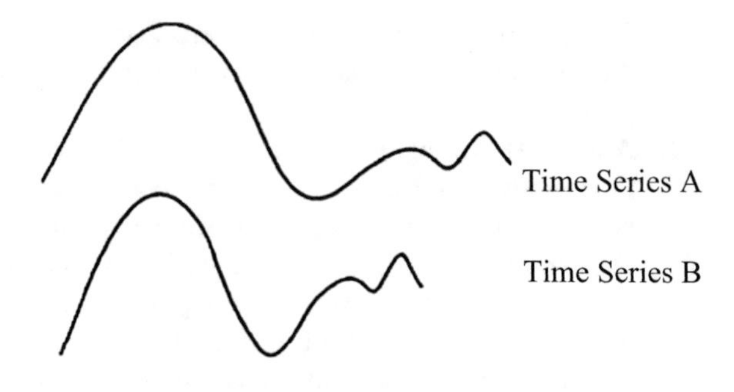

Fig. 4.14 Linear drift of time series

$$X' = X + f(t), \quad t = 1, 2, \ldots, n \tag{4.8}$$

In this expression, *f*(*t*) is a time related function which indicates the reason for linear drift.

(5) Time Scaling. As shown in Fig. 4.14, time scaling occurs when two time series are similar in wave form but their widths scale in the same proportion.
(6) Time Warping. As shown in Fig. 4.15, time warping occurs when two time series are similar in waveform but their peaks and troughs deviate a little rather than strictly align. Time scaling is a special case of time warping.

In order to remove the effect of amplitude shifting and scaling on similarity measurement, a normalized preprocessing of time-series data is necessary, i.e., horizontal base line is set at 0, and amplitude at 1. Two methods, i.e. conventional method and statistical method, are used for normalization of preprocessing. For time series $\boldsymbol{X} = \{x_1, x_2, \ldots, x_n\}$, the application of the conventional method leads to the mathematic expression below:

$$\operatorname{norm}(x_i) = \frac{x_i - \min(\boldsymbol{X})}{\max(\boldsymbol{X}) - \min(\boldsymbol{X})} \tag{4.9}$$

Suppose the mean value and variance of time series *X* are respectively $\mu(\boldsymbol{X})$ and $\rho(\boldsymbol{X})$, the following statistical formula can then be obtained:

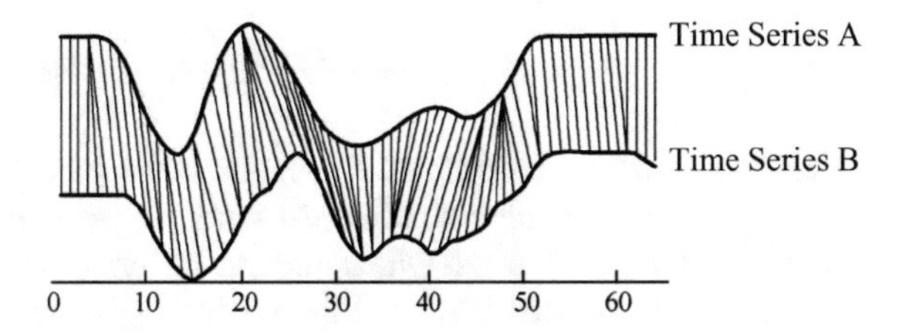

Fig. 4.15 Warping scaling of time series

$$\text{norm}(x_i) = \frac{x_i - \mu(\boldsymbol{X})}{\rho(\boldsymbol{X})} \quad (4.10)$$

The current research on method of similarity measurement of time series mainly covers ordinary norm distance, DWT of dynamic time warping distance, pattern distance D_M, and so on. Many research efforts can be seen as expansions of the methods just mentioned.

1. Ordinary Norm Distance

Suppose time series $\boldsymbol{A} = \{a_1, a_2, \ldots, a_n\}$ and $\boldsymbol{B} = \{b_1, b_2, \ldots, b_n\}$ are of equal length n, then the distance between them can be calculated by means of the following equation, L_p in which is the ordinary norm distance, or Minkowski Distance:

$$L_p(\boldsymbol{A}, \boldsymbol{B}) = \left(\sum_{i=1}^{n} |a_i - b_i|^p \right)^{1/p}, \quad 1 \le p \le \infty \quad (4.11)$$

When p = 1, L_1 is referred to as Manhattan Distance; when p = 2, L_2 is actually Euclidean Distance, which is expressed as follows:

$$L_2(\boldsymbol{A}, \boldsymbol{B}) = \sqrt{\sum_{i=1}^{n} (a_i - b_i)^2} \quad (4.12)$$

Euclidean Distance is also referred to as Euclidean norm, and can be represented in form of linear algebra as follows:

$$L_2(\boldsymbol{A}, \boldsymbol{B}) = \sqrt{(\boldsymbol{A} - \boldsymbol{B})^{\mathrm{T}} \sum (\boldsymbol{A} - \boldsymbol{B})} \quad (4.13)$$

In this expression, $\sum$ is a diagonal matrix while T a transposition operator.

When $p = \infty$ occurs, L_∞ indicates the maximum distance between two time series in a pair, and is represented in the equation below:

$$L_p(A, B) = \max_{i=1}^{n}\{|a_i - b_i|\} \tag{4.14}$$

In this equation, an optimal $\boldsymbol{L}_1$ results when measuring error satisfies additive Laplace distribution, and therefore, $\boldsymbol{L}_1$ works better where there is impulsive noise. An optimal $\boldsymbol{L}_2$ will result when measuring error satisfies additive independent identical Gaussian distribution; this is why $\boldsymbol{L}_2$ is currently the most widely used method. Elsewhere in the discussion of similarity measurement of time series, there have also been L_∞-based similarity models to determine whether two time series are similar or not.

Keogh and others, through introducing into time series processing the notion of weighting, expand Euclidean Distance to Weighted Euclidean Distance. In doing so, they assign a weight value to the points of the time series according to their importance. For example, if a weight value is assigned to each point in $A = \{a_1, a_2, \ldots, a_n\}$, it can be represented as $\boldsymbol{A} = \{a_1, a_2, \ldots, a_n | w_1, w_2, \ldots, w_n\}$; consequently the equation below is obtained:

$$L_2[(\boldsymbol{A}, \boldsymbol{W}), \boldsymbol{B}] = \sqrt{\sum_{i=1}^{n} w_i (a_i - b_i)^2} \tag{4.15}$$

If Eq. (4.15) is applied to $\boldsymbol{L}_p$, a weighted $\boldsymbol{L}_p$ will be obtained in the equation below:

$$L_p[(\boldsymbol{A}, \boldsymbol{W}), \boldsymbol{B}] = \left(\sum_{i=1}^{n} w_i |a_i - b_i|^p\right)^{1/p} \tag{4.16}$$

As an ordinary norm distance, $\boldsymbol{L}_p$ is characterized with the following four attributes:

(1) Self-similarity: $\boldsymbol{L}_2(\boldsymbol{A}, \boldsymbol{B}) = 0$, when and only when $\boldsymbol{A} = \boldsymbol{B}$.
(2) Nonnegativity: $L_2(\boldsymbol{A}, \boldsymbol{B}) \geq 0, \forall \boldsymbol{A}, \boldsymbol{B}$.
(3) Symmetry: $L_2(\boldsymbol{A}, \boldsymbol{B}) = L_2(\boldsymbol{B}, \boldsymbol{A})$.
(4) Triangle inequality: $L_2(\boldsymbol{A}, \boldsymbol{B}) \leq L_2(\boldsymbol{A}, \boldsymbol{C}) + L_2(\boldsymbol{B}, \boldsymbol{C})$.

It can be seen that the triangle inequality of $\boldsymbol{L}_p$ makes it possible to do precise search by building a query index. In distance calculating of time series in database, change of one of the time series will make little difference to others in their distance calculating and indexing. In the search space, any object series that surpasses the preset threshold distance will be deleted so that the unqualified index nodes are quickly filtered. As a result of this, quick search is realized. This quick search strategy has been used such multidimensional index models as Range Query and Nearest Neighbor Query, R-Tree, R*-Tree, and M-Tree for example. Figure 4.16 is a diagram of calculation using ordinary norms.

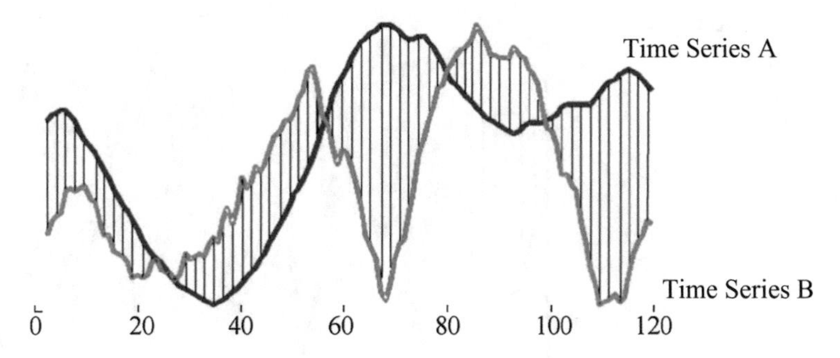

Fig. 4.16 Calculation using ordinary norms

On the basis of $\boldsymbol{L_p}$, ordinary norm distance, Sangjun Lee, Dongseop Kwon and Sukho Lee propose the notion of Minimum Distance, which is marked as $L_p^{\min}$ and defined in terms of a threshold value ε and time series **A** and **B**. If time series **A** and **B** are of equal length and satisfy the equation below, in which $m = \sum_{i=1}^{n} (a_i - b_i)/n$, they can be considered as similar in shape:

$$L_p^{\min}(\boldsymbol{A}, \boldsymbol{B}) = \left(\sum_{i=1}^{n} |a_i - b_i - m|^p \right)^{1/p} \leq \varepsilon \tag{4.17}$$

If similarity is to be determined in terms of shape, and consideration of the vertical position of time series is not necessary, this definition may serve as a highly adaptable model.

2. Dynamic Time Warping (DTW) Distance

Since calculation of $\boldsymbol{L_p}$ requires a strict point-to-point correspondence between two time series, $\boldsymbol{L_p}$ almost does not work in the case of scaling and warping of time axis although much research effort has been devoted to its improvement. The problem is solved by Berndt and Clifford, who introduce into similarity query of time series DTW distance, a method widely used in speech recognition. DTW distance works very well for similarity measurement when scaling and warping of time axis occurs.

Different from $\boldsymbol{L_p}$, DTW distance does not require a point-to-point mach between two time series. When two time series are different in length, DTW distance allows a self-replication of the points, which are then aligned for comparison. When warping of time axis occurs, the points at the warping will be self-replicated so that similar wave forms of two time series many be aligned and compared. As indicated in Fig. 4.17, the points marked with small circles are the self-replicated points.

For two time series $\boldsymbol{A} = \{a_1, a_2, \ldots, a_n\}$ and $\boldsymbol{B} = \{b_1, b_2, \ldots, b_n\}$ which are n and m in length, and $n \neq m$, construct in chronological order a matrix $\boldsymbol{M}$ which represents the point-to-point correspondence between $\boldsymbol{A}$ and $\boldsymbol{B}$. As shown in Fig. 4.18, matrix element $\boldsymbol{M}_{ij}$ indicates the distance between a_i and b_j, which is

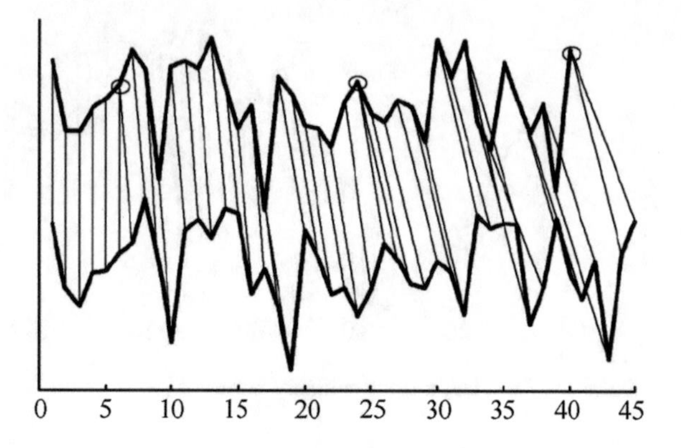

Fig. 4.17 Calculation of DTW distance

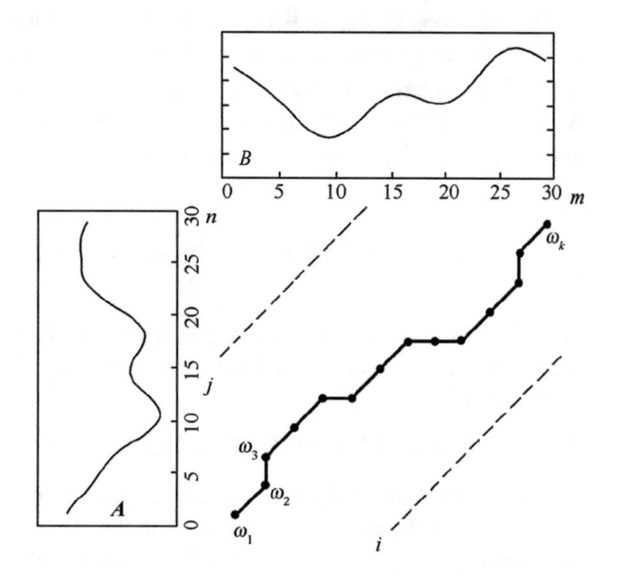

Fig. 4.18 DTW path of time series

written as d (a_i, b_j). Also in Fig. 4.18, the broken line is the time warping path $\boldsymbol{W}$, indicating point-to-point mapping relation between time series $\boldsymbol{A}$ and $\boldsymbol{B}$; the kth matrix element is written as $w_k = (i, j)$, and therefore, the equation below can be obtained:

$$\boldsymbol{W} = (w_1, w_2, \ldots, w_k), \quad \max(n, m) \leq k \leq n + m - 1 \tag{4.18}$$

DTW path $\boldsymbol{W}$ is supposed to meet the following requirements:

(1) Boundary condition which requires that the DTW path start as the lower left corner of the matrix and end at the upper right corner, that is $w_1 = (1, 1)$, $w_k = (n, m)$.
(2) Continuity, which requires that the step numbers allowed by the DTW path should exclude the adjacent elements, including the diagonally adjacent ones.

That is, if $w_k = (a, b)$, then $w_{k-1} = (a', b')$, in which $a - a' \leq 1$ and $b - b' \leq 1$.

(3) Monotonicity, according which the points on the DTW path must be chronically monotonous. That is, if $w_k = (a, b)$, then $w_{k-1} = (a', b')$, in which $a - a' \geq 0$ and $b - b' \geq 0$.

Among the DTW paths that meet the requirements, the minimum warping path is the optimal path. And therefore, a further requirement is proposed as follows:

$$\mathrm{DWT}(\boldsymbol{A}, \boldsymbol{B}) = \min\left\{\sqrt{\sum_{k=1}^{K} w_k / K}\right. \tag{4.19}$$

In this equation, since DTW path may vary in length, denominator K may be used for compensation.

In summary, DWT distance can be defined as follows:

$$\mathrm{DWT}(\boldsymbol{A}, \boldsymbol{B}) = \begin{cases} \infty, & \text{if } n = 0 \text{ or } m = 0 \\ d(a_1, b_1) + \min\begin{cases} \mathrm{DTW}[\mathrm{Rest}(\boldsymbol{A}), \mathrm{Rest}(\boldsymbol{B})] \\ \mathrm{DTW}[\mathrm{Rest}(\boldsymbol{A}), \boldsymbol{B}] \\ \mathrm{DTW}[\boldsymbol{A}, \mathrm{Rest}(\boldsymbol{B})] \end{cases} & \end{cases} \tag{4.20}$$

In this expression: $d(a_1, b_1) = |a_1 - b_1|$; Rest($\boldsymbol{A}$), analogously represented as $\{(t_2, a_2), \ldots, (t_n, a_n)\}$, refers to the rest elements in time series A after the first element is removed.

However, since the definition above does not explicate how to calculate DTW distance between two time series, the distance can be obtained through dynamic programming method based on a cumulative distance matrix. A cumulative distance matrix is in fact a representation of recurrent relations; for time series A and B, defined for the present study, their cumulative distance can be defined as follows:

$$r(i, j) = d(a_i, y_j) + \min\{r(i-1, j), r(i, j-1), r(i-1, j-1)\} \tag{4.21}$$

DTW distance defines the optimum matching relationship between two time series; in addition to its support to warping of time axis, it can also be used in length measuring of time series. However, in the case of high time complexity ($\boldsymbol{O}(nm)$), which may even be as high as $\boldsymbol{O}(n^2\boldsymbol{L})$ in subseries matching, DTW distance is not suitable for data mining of mass time-series data. In $\boldsymbol{O}(n^2\boldsymbol{L})$, $\boldsymbol{L}$ is the length of subseries.

3. Pattern distance

Either ordinary norm distance or DTW distance is a point-to-point distance, that is, the distance between two points should be calculated. But distance of this type is characterized with an inherent weakness when time series matching is done in terms of change trend. This is because point-to-point distance is a type of static measuring

Fig. 4.19 PLR of time series

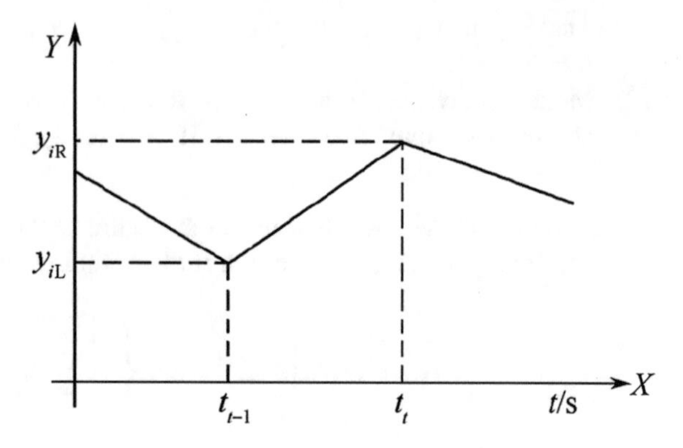

and does not work well for time series, dynamic in nature. In terms of the change trend of time series, and on the basis of PLR (piecewise linear representation), pattern distance divides time series into a number of subsets with each representing a pattern. In this way, pattern distance not only works well in measuring the similarity between two time series in terms of change trend, but also enables distance measuring under different resolutions.

By means of PLR and as shown in Fig. 4.19, a time series is represented as several neighboring straight lines expressed as follows:

$$\boldsymbol{S} = \{(y_{1\mathrm{L}}, y_{1\mathrm{R}}, t_1), (y_{i\mathrm{L}}, y_{i\mathrm{R}}, t_i), \ldots, (y_{K\mathrm{L}}, y_{K\mathrm{R}}, t_K)\}$$

In this equation, $y_{i\mathrm{L}}$ and $y_{i\mathrm{R}}$ represent respectively the initial (left end) and final (right end) value of the ith straight-line segment; t_i is the time when the ith segment stops; $\boldsymbol{K}$ is the number of segments a whole piece of time series is segmented into, and the selection of $\boldsymbol{K}$ can be calculated by using the from bottom to top algorithm proposed by Keogh.

Pattern represents the change trend of a subset of time series. A pattern normally takes the form of a three-factor set, namely, {Rising, Holding, Falling}, expressed as $\boldsymbol{M} = \{1, 0, -1\}$ for a convenient calculation. $M \in \boldsymbol{M}$ represents a factor in the pattern as a subset; for a time series $\boldsymbol{S}$, its pattern can be expressed in terms of a pattern-time pair as follows:

$$\boldsymbol{S} = \{(m_1, t_1), \ldots, (m_i, t_i), \ldots, (m_n, t_n)\}$$

As is indicated in this equation, $m_i \in \boldsymbol{M}$, $i = 1, 2, \ldots, \boldsymbol{n}$; $\boldsymbol{t}_1, \boldsymbol{t}_2, \ldots, \boldsymbol{t}_n$ represents the time when the pattern comes to an end; and $\boldsymbol{n}$ is the number of patterns that time series $\boldsymbol{S}$ is segmented into.

Pattern distance refers to the distance between two patterns with the same hold time, that is, for the kth and pth patterns, there should be $\boldsymbol{D}_M(s_k, s_p) = |m_k - m_p|$, in which $s_k = (m_k, t_k)$, $s_p = (m_p, t_p)$ and $\boldsymbol{D}_M \in \{0, 1, 2\}$; and, when and only when $m_k = m_p$, $\boldsymbol{D}_M = 0$.

In its application to time series research, pattern distance can be used as an indicator of the degree of difference in change trend between two time series of equal length. S_1 and S_2, for instance, are expressions of patterns of two equal length time series to be matched:

$$S_1 = \{(m_1, t_1), \ldots, (m_{1i}, t_{1i}), \ldots, (m_{1N}, t_{1N})\}$$

$$S_2 = \{(m_2, t_2), \ldots, (m_{2i}, t_{2i}), \ldots, (m_{2M}, t_{2M})\}$$

In the expressions above, $t_{1N} = t_{2M}$, S_{1i} and S_{2j} represent respectively the *i*th and *j*th pattern in S_1 and S_2.

For a random pattern of S_1 or S_2, its hold time is expressed as: $t_{1ih} = t_{1i} - t_{1(i-1)}$ and $t_{2jh} = t_{2j} - t_{2(j-1)}$. When $t_{1ih} = t_{2jh}$, if $M = N$, pattern distance between S_1 and S_2 is:

$$D_M(s_1, s_2) = \frac{1}{t_N}\sum_{i=1}^{N} t_{ih} D_M(S_{1i}, S_{2i}), \quad \text{and} \quad \frac{1}{t_N}\sum_{i=1}^{N} t_{ih} = 1 \tag{4.22}$$

In this expression, t_N is the length of time series; t_{ih}, the hold time of the ***i***th pattern; and ***N***, the number of segmented patterns. t_{ih} works to assign a weighting to different hold times; the longer the hold time is, the greater proportion the corresponding pattern will take among all the patterns.

4.2 Similarity Search Method in Time Series Based on Slope Distance

As has been introduced in the previous section, three similarity measuring methods, namely ordinary norm distance, DTW distance and pattern distance, can work effectively in different application backgrounds. Among the three methods, pattern distance is closer to natural language description with well defined physical significance and suitable segmentation. Compared with the other two methods, however, pattern distance is still a crude method for its unclear conclusions. For a making up, slope distance (SD) method for measuring time series similarity will be introduced in this section.

4.2.1 Slope Set Representation of Time Series

1. Slope Set

PLR of time series S can be expressed as $S = \{(x_0, x_1, t_1), \ldots, (x_{i-1}, x_i, t_i), \ldots, (x_{n-1}, x_n, t_n)\}$, in which x_{i-1} and $x_i (i = 1, 2, \ldots, n)$ represent, respectively, the initial and final end value of the *i*th straight-line segment, t_i the time when the *i*th straight-line segment comes to an end, and $n - 1$, the number of straight-line segment time series *S* is segmented into. As to the rising, holding and falling of the straight-line segments, they can be represented in terms of slope, that is, time series ***S*** can always be represented as a set of line segments with a slope.

Definition 4.1 Suppose k_i is the slope of the piecewise straight lines, and then *k*, which is a set of all k_i's, is the slope set of. In terms of *k*, time series ***S*** can be expressed as follows:

$$\boldsymbol{S} = \{(k_0, t_1), \ldots, (k_{i-1}, t_i), \ldots, (k_{n-1}, t_n)\} \tag{4.23}$$

In this expression, $i = 1, 2, \ldots, n$; $t_1, t_2, \ldots, t_n$ represent the time when a line segment stops; and $n - 1$ is the number of the segments.

2. Algorithm of Slope Set Search

This algorithm transforms PLR into the slope of the time series, which will result in $(n - 1)$ time series represented in terms of slope. The algorithm goes like this:

```
input: time series S expressed in terms of PLR
output: time series S' expressed in terms of k
result = {};
for i = 1: n
{

k = (x_i − x_i-1) / (t_i − t_i-1);
result( )= ;

}
return result;
```

As has been shown, this algorithm scans the time series database only one time and features high computational efficiency.

4.2.2 *Slope Distance Measurement for Time Series*

Suppose ***S***′ and ***S***″ are two time series of equal length, and expressed in terms of slope:

$$\boldsymbol{S}' = \{(k'_0, t'_1), \ldots, (k'_{i-1}, t'_i), \ldots, (k'_{n-1}, t'_n)\} \tag{4.24}$$

$$\boldsymbol{S}'' = \{(k''_0, t''_1), \ldots, (k''_{i-1}, t''_i), \ldots, (k''_{n-1}, t''_n)\} \tag{4.25}$$

Definition 4.2 From time series $\boldsymbol{f} = \{x_1, x_2, \ldots, x_n\}$, extract three continuous data x_{i-2}, x_{i-1} and x_i, and let recursive increment be $d_{x0} = x_{i-1} - x_{i-2}$ and $d_{x1} = x_i - x_{i-1}$.

Then let the deviation coefficient be β; l refers to times of occurrence of the elements in time series f that satisfy the condition $|d_{x1}| > \beta|d_{x0}|$, and serves as a counter of small increment deviation. And then let α be the maximum allowable deviation value; when $l \geq \alpha$, x_i is seen as an outlier.

Definition 4.3 In view of the direction of slope, a self-adaptive parameter γ is introduced, and is called a direction coefficient. γ is defined as follows:

$$\gamma = \begin{cases} 1 & \text{if } \left(k'_{i-1} \times k''_{i-1}\right) \geq 0 \\ C & \text{if } \left(k'_{i-1} \times k''_{i-1}\right) < 0 \end{cases} \tag{4.26}$$

In this definition, $C > 1$.

Definition 4.4 On the basis of S' and S'' value obtained from Eqs. 4.24 and 4.25, set the original slope distance of S' and S'' as follows:

$$D_K(S', S'') = \left| \sum_{i=2}^{n} (t_i - t_{i-1})(k'_{i-1} - k''_{i-1})/t_n \right| \tag{4.27}$$

In this expression, $t_i - t_{i-1} = \max\left\{(t'_i - t'_{i-1}), (t''_i - t''_{i-1})\right\}$ and $t_n = \max\left\{t'_n, t''_n\right\}$. On the basis of Definitions 4.1, 4.2 and 4.3, let the SD of S' and S'' be:

$$D_K(S', S'') = \begin{cases} \left| \sum_{i=2}^{n} \gamma(t_i - t_{i-1})(k'_{i-1} - k''_{i-1})/t_n \right|, & l < \alpha \\ \infty, & l \geq \alpha \end{cases} \tag{4.28}$$

In this expression, t_n refers to the total time length; $(t_i - t_{i-1})$ serves as a weight, the longer the time, the greater the weight, and at the same time, $\left(\sum_{i=2}^{n} (t_i - t_{i-1})\right)/t_n = 1$, and $i = 2, 3, \ldots, n$.

A review of the above definitions shows that SD of this type has its clear physical meaning, and is therefore closer to people's intuitive judgment. Meanwhile, direction coefficient together with counter of small increment deviation results in higher computational efficiency and more realistic results. Moreover, distance calculation by means of slope reduces much calculating time since it makes it possible to go without the normalized procedures previously necessary in processing data of different types and units of measurements.

4.2.3 Verification on Clustering of Flight Data Based on Slope Distance

For the discussion in this section, four types of parameters have been extracted from the flight data of an aircraft in a sortie. The parameters indicate operation of HP and LP rotors in left and right engines with each parameter curve consisting of 746

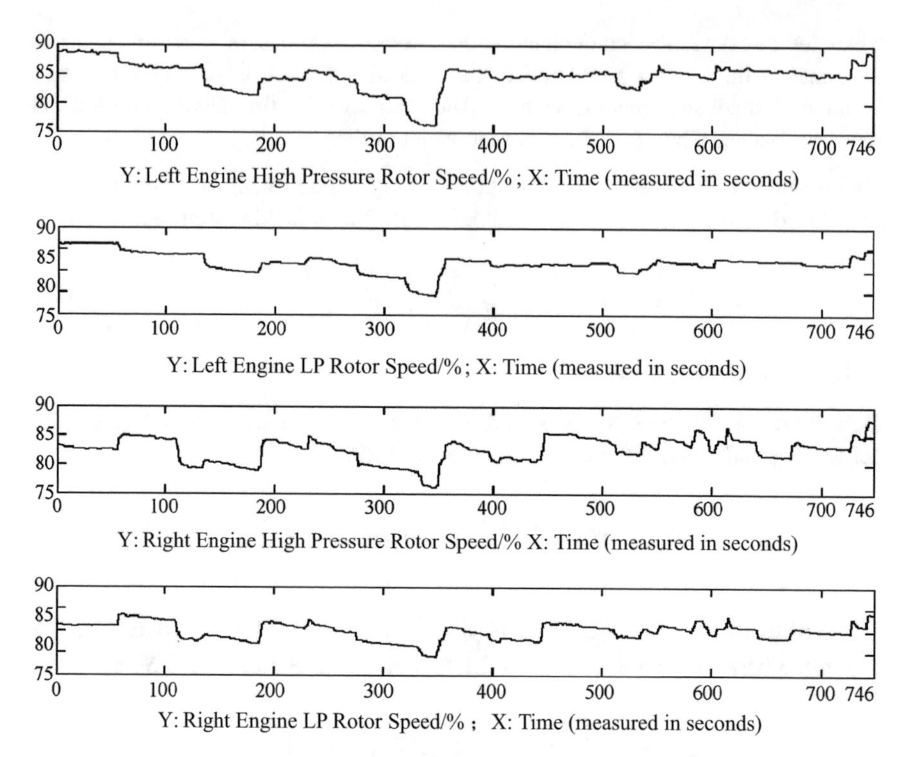

Fig. 4.20 Speed-time curves: left/right engine HP/LP rotor speed of a sortie

discrete points. By means of slope representation and PLR, the four parameter curves are divided into 20 line segments as shown in Fig. 4.20. In a top-down order, Fig. 4.20 displays four time-varying curves, namely, n_{2l}, HP rotor speed of left engine; n_{1l}, LP rotor speed of left engine; n_{2r}, HP rotor speed of right engine; and n_{1r}, LP rotor speed of right engine. The visual features of the curves show similarity between n_{2l} and n_{2r}, and between n_{1l} and n_{1r}.

1. Calculation of Euclidean Distance

Calculation results of Euclidean distance are listed in Table 4.1, which indicates that Euclidean distance fails to work out valid results in distinguishing n_{2l} and n_{2r}. Although minimum distance occurs between them, they are not of the same type.

2. Calculation of Original Slope Distance

Calculation results of original slope distance are listed in Table 4.2, in which n_{2l}, n_{1l} and n_{2r} are classified as one type while n_{1r} as another. The reason for this classification is that if Eq. 4.27 is used when slope values of the curves stay near to 0, plus or minus signs are neglected by the algorithm. Consequently, two curves with opposite trends turn out to be similar ones. Hierarchical classification of the four curves is shown in Fig. 4.21.

Table 4.1 Calculation results of Euclidean distance

	n_{21}	n_{11}	n_{2r}	n_{1r}
n_{21}	0	61.42	49.15	70.57
n_{11}		0	98.82	85.02
n_{2r}			0	53.37
n_{1r}				0

Table 4.2 Calculation results of original slope distance

	n_{21}	n_{11}	n_{2r}	n_{1r}
n_{21}	0	0.0491	0.0859	0.1269
n_{11}		0	0.0946	0.1419
n_{2r}			0	0.1152
n_{1r}				0

3. Calculation of Slope Distance

In Table 4.3 are listed the results of slope distance calculation, in which Eq. 4.28 is used and $\boldsymbol{C} = 3.5$. As indicated by the table, two pairs of similar curves can be identified, namely n_{21} and n_{2r}, n_{11} and n_{1r}, which agree to the real situation. Hierarchical Classification of the four curves is shown in Fig. 4.22.

With the introduction into calculation small increment deviation counter and direction coefficient, the slope-based measuring method has such advantages as clearer physical conception, insensitivity to data size, and more emphasis on similarity of curve patterns. This method goes without the normalized procedures in processing data of different types and units of measurement, and therefore costs much less calculation time. Owing to the use of PLR method, both pattern and slope distance display a character of multi-resolution, and may provide data mining with a new approach. But it should be noted that in its engineering application, threshold value of distance can only be determined in experiments.

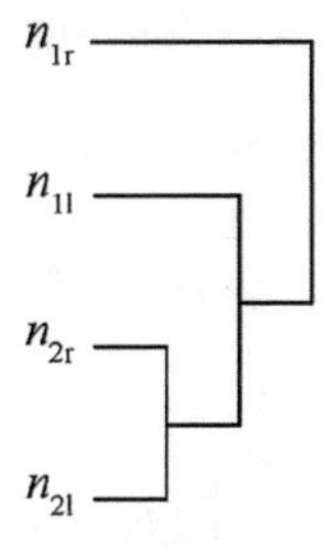

Fig. 4.21 Hierarchical classification based on original slope distance

Table 4.3 Calculation results of improved slope distance

	n_{21}	n_{11}	n_{2r}	n_{1r}
n_{21}	0	0.9418	0.4114	0.9413
n_{11}		0	0.7657	0.5640
n_{2r}			0	0.6249
n_{11}				0

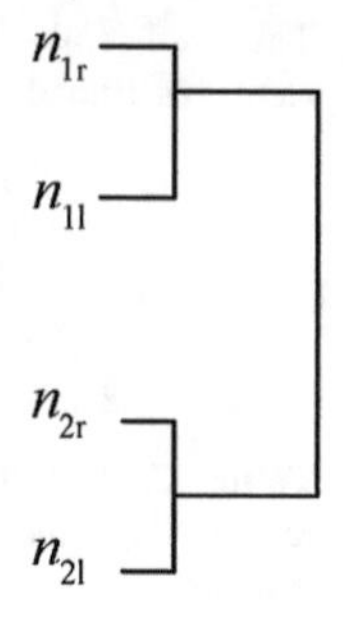

Fig. 4.22 Hierarchical classification based on improved slope distance

4.3 Similarity Search Method in Time Series Based on Included Angle Distance

The discussion in the previous section shows that for similarity search of time series, slope-distance-based method incorporates into itself small increment deviation counter, direction coefficient, and the like self-defined parameters. Although this method is effective in eliminating misjudgment of similarity in the case of sudden change of slope, its engineering application is limited by high dependence of the parameters on human experience and data sources. In order to lessen the dependence on experience and data sources, an angular distance related search method will be proposed in this section.

4.3.1 Representation of Included Angle Set for Time Series

1. PLR of Time Series

PLR (piecewise linear representation) is an important method of feature extraction and dimension reduction in time series research. Practically in ordinary database processing, PLR approximation method proves to be as good as traditional DFT. Let $\boldsymbol{N}$ be the length of time series $\boldsymbol{X}$, whose APLR (adaptive piecewise linear representation) is its PLR plus automatic determination of segment amount. Thus time series $\boldsymbol{X}$ can be expressed as follows:

$$\boldsymbol{X} = \{(x_1, x_2, t_2), \ldots, (x_{i-1}, x_i, t_i), \ldots, (x_{n-1}, x_n, t_n)\} \tag{4.29}$$

In this expression, x_{i-1} and x_i, respectively, refer to the initial and final value of the ith straight-line segment; t_i is the time when the ith segment stops; and n stands for the amount of straight-line segments of the time series. APLR of $\boldsymbol{X}$ can be done by means of extracting edge points according to the slope, a method featuring small fitting error and low computational complexity.

2. Angle Set of Time Series

Definition 4.5 PLR of time series $\boldsymbol{X}$ can be expressed as follows:

$$\boldsymbol{X} = \{(x_1, x_2, t_2), \ldots, (x_{i-1}, x_i, t_i), \ldots, (x_{n-1}, x_n, t_n)\}$$

In terms of a set of included angle of adjacent line segments, $\boldsymbol{X}$ can be further and approximately expressed as:

$$\boldsymbol{X} = \{\alpha_1, \alpha_2, \ldots, \alpha_{n-1}\} \tag{4.30}$$

In this expression, α_i is the anticlockwise included angles between adjacent line segments, that is (x_i, x_{i+1}, t_{i+1}) and $(x_{i+1}, x_{i+2}, t_{i+2})$. Then let the clockwise included angle between adjacent line segments be β_i, value of α_i can be obtained from the expression below:

$$\begin{cases} \alpha_i = \arccos\frac{a^2+b^2-c^2}{2ab}, & \alpha_i \le \beta_i \\ \alpha_i = 2\pi - \arccos\frac{a^2+b^2-c^2}{2ab}, & \alpha_i > \beta_i \end{cases} \tag{4.31}$$

As shown by the expression above, $\boldsymbol{R}^N \to \boldsymbol{R}^{n-1}$ projection is realized through angle set representation of time series. Since $n - 1 \ll N$ normally occurs, $\boldsymbol{R}^N \to \boldsymbol{R}^{n-1}$ projection will results in dimension reduction of time series.

3. Search Algorithm of Angle Set

For a time series with n linear segments, search algorithm of angle set can be used to represent it in terms of angles, that is, this time series will be transformed into $n - 1$ time series in the form of a series of angles. The following is the description of the working of the algorithm:

Input: APLA of time series X;
Output: time series $\boldsymbol{X}'$ represented as α_i;
begin
result = { };
for i = 1: n − 1
{

If $\alpha_i \le \beta_i$ then $\alpha_i = \arccos\frac{a^2+b^2-c^2}{2ab}$;
if $\alpha_i > \beta_i$ then $\alpha_i = 2\pi \arccos\frac{a^2+b^2-c^2}{2ab}$;
result (i) = α_i;

}
$\boldsymbol{X}'$ = result;
return $\boldsymbol{X}'$;
end

4.3.2 *Included Angle Distance Measurement for Time Series*

1. Time Alignment of Angle Set

Normally, after PLR is applied to two time series, the endpoints, which are borders of line segments, may not fully match in time; and similarly, the segments to be compared with endpoints as their borders may not be of equal length either. Therefore, for an effective similarity search, time alignment should be done to the two sets of angels.

Definition 4.6 Time alignment is redivision of paired time intervals for comparison by repositioning endpoints so that they are of equal length and meet the requirements of the algorithm. As stipulated in the algorithm, π is the included angle between adjacent line segments at the aligned time point.

For example, if time alignment is done to two time series $\boldsymbol{X}' = \{(\alpha_1', t_3), (\alpha_2', t_4)\}$ and $\boldsymbol{X}'' = \{(\alpha_1'', t_2), (\alpha_2'', t_5)\}$, they can be represented as follows:

$$X' = \{(\alpha_1', t_2), (\alpha_2', t_3), (\alpha_3', t_4), (\alpha_4', t_5)\}$$

$$X'' = \{(\alpha_1'', t_2), (\alpha_2'', t_3), (\alpha_3'', t_4), (\alpha_4'', t_5)\}$$

In these expressions, $\alpha_1' = \alpha_4' = \alpha_2'' = \alpha_3'' = \pi$.

The effect of time alignment is shown in Fig. 4.23.

2. Definition and Nature of Angular Distance

Definition 4.7 As shown below, $\boldsymbol{X}'$ and $\boldsymbol{X}''$ are time-aligned time series of equal length and represented in terms of a set of angles.

$$\boldsymbol{X}' = \left\{\alpha_1', \alpha_2', \ldots, \alpha_{n-1}'\right\}$$

$$\boldsymbol{X}'' = \left\{\alpha_1'', \alpha_2'', \ldots, \alpha_{n-1}''\right\}$$

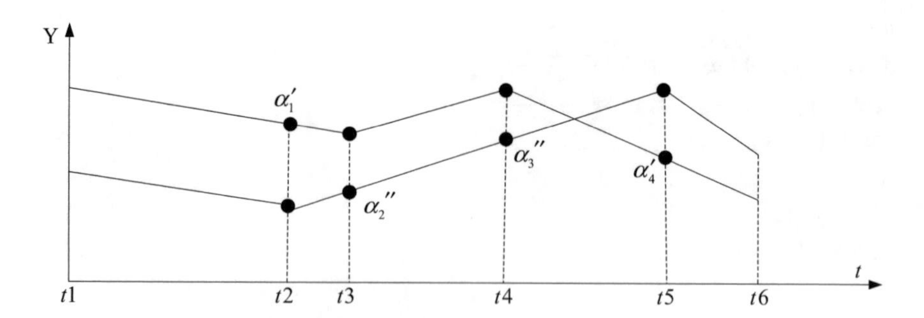

Fig. 4.23 Time alignment of time series

The angular distance, which is a difference measurement, between $\boldsymbol{X}'$ and $\boldsymbol{X}''$ can be expressed as follows:

$$D_a(\boldsymbol{X}', \boldsymbol{X}'') = \frac{\sum_{i=1}^{n-1} |\alpha_i' - \alpha_i''|}{\pi(n-1)} \tag{4.32}$$

Theorem 4.1 *Following attributes can be summarized for time series similarity measured in terms of angular distance:*

(1) Nonnegativity: $D_a(\boldsymbol{X}', \boldsymbol{X}'') > 0 \quad \text{if } \boldsymbol{X}' \neq \boldsymbol{X}''$.
(2) Self-similarity: $D_a(\boldsymbol{X}', \boldsymbol{X}'') = 0 \quad \forall \text{ if } \quad \boldsymbol{X}' = \boldsymbol{X}''$.
(3) Symmetry: $D_a(\boldsymbol{X}', \boldsymbol{X}'') = D_a(\boldsymbol{X}', \boldsymbol{X}'') \quad \forall\, \boldsymbol{X}', \boldsymbol{X}''$.
(4) Triangle inequality: $D_a(\boldsymbol{X}', \boldsymbol{X}'') \leq D_a(\boldsymbol{X}', \boldsymbol{X}''') + D_a(\boldsymbol{X}', \boldsymbol{X}''')$.
(5) Roundedness: $D_a(\boldsymbol{X}', \boldsymbol{X}''') \in [0, 1]$.

According to Definitions 4.6 and 4.7, attributes (1), (2), and (3) are obviously valid.

Attribute (4) guarantees that the proposed similarity search algorithm based on angular distance satisfies the condition of triangle inequality. It can also be used to build an index for quick search. When distance calculation is done to time series in database, change of the time series will have very small effect on distance calculation of other series, and on index query. In the search space, any object series that exceeds the preset threshold value will be removed, so that unqualified index nodes can be quickly filtered out for a quick search.

Attribute (5) indicates that if the angular distance between two series is 0, it can be concluded that they are fully identical in shape; that if the angular distance between two series is 1, it can be concluded that they are completely different in shape; and that the open interval (0, 1) not only reveals the level of similarity between the series, but also that of the operation conditions reflected by the data with different change trends.

As shown in the discussion above, included angle distance is a shape-based similarity measurement of time series. This measurement method is characterized with following features:

(1) Invariance. Since similarity is measured in terms of angular difference between adjacent line segments, this method overcomes inconsistent description caused by translation and rotation.
(2) High robustness. Since angular distance is defined on the basis of PLR, it is insensitive to local disturbance.
(3) High computational efficiency. Since this method satisfies the condition of triangle of inequality, it helps to establish a search index for quick query.

3. Similarity Search Method Based on Included Angle Distance

Similarity search method by means of angular distance identifies in a time series database the series that is considered similar to the given pattern. That is, in database $\mathbf{X}_1, X_2, \ldots, X_n$, and for a standard series X and similarity threshold value ε, search Xi that satisfies the condition of $D_a(X, X_i) \leq \varepsilon$, all series being of equal length N.

The algorithm goes as follows:

Step 1: According to the proposed algorithm, apply dimension reduction to query series X and time series database $\mathbf{X}_1, X_2, \ldots, X_n$, transforming them into series of angle sets represented as X' and $X'_1, X'_2, \ldots, X'_n$.

Step 2: Apply time alignment to X' and $\mathbf{X}_1, X_2, \ldots, X_n$, with the time-aligned series still represented as X' and $X'_1, X'_2, \ldots, X'_n$.

Step 3: On the basis of included angle distance and similarity threshold, store and search on R tree candidate series that are similar to series X'.

Step 4: Calculate the actual distance between standard series X and candidate series corresponding to the original ones, then obtain the final results by filtering out unqualified series.

Included angle-based algorithm of similarity search incorporates into itself PLR, angular representation of time series, and time alignment. From the time complexity of the algorithms mentioned above, that is, $(N\lg N)$ for DFT and $O(N)$ for DWT, it can be seen that PLR, time complexity of which is also $O(N)$, works better than DFT, and is as good as DWT. Moreover, since angular representation and time alignment can be done simultaneously, PLR reduces times of database scanning and therefore improves computational efficiency.

4.3.3 Verification on Clustering of Time Series Based on Included Angle Distance

In this section, three methods of similar pattern identification will be used for a comparative purpose. In the methods, three types of distance between time series are calculated respectively, namely Euclidean distance D_E, slope distance D_K, and included angle distance D_a.

Flight data of an aircraft in a sortie are used as samples, from which data over a certain period of time are selected. The selected data are actually four time series, each consisting of 746 discrete points. As shown in Fig. 4.24, four speed-time curves are provided here in a top-down order: n_{2l}, HP rotor speed of the left engine; n_{1l}, LP rotor speed of left engine; n_{2r}, HP rotor speed of right engine; and n_{1r}, LP rotor speed of right engine. Also in Fig. 4.24, the vertical ordinate is for parameters managed for the present purpose, while the horizontal ordinate for time (measured in seconds) of flight data recording.

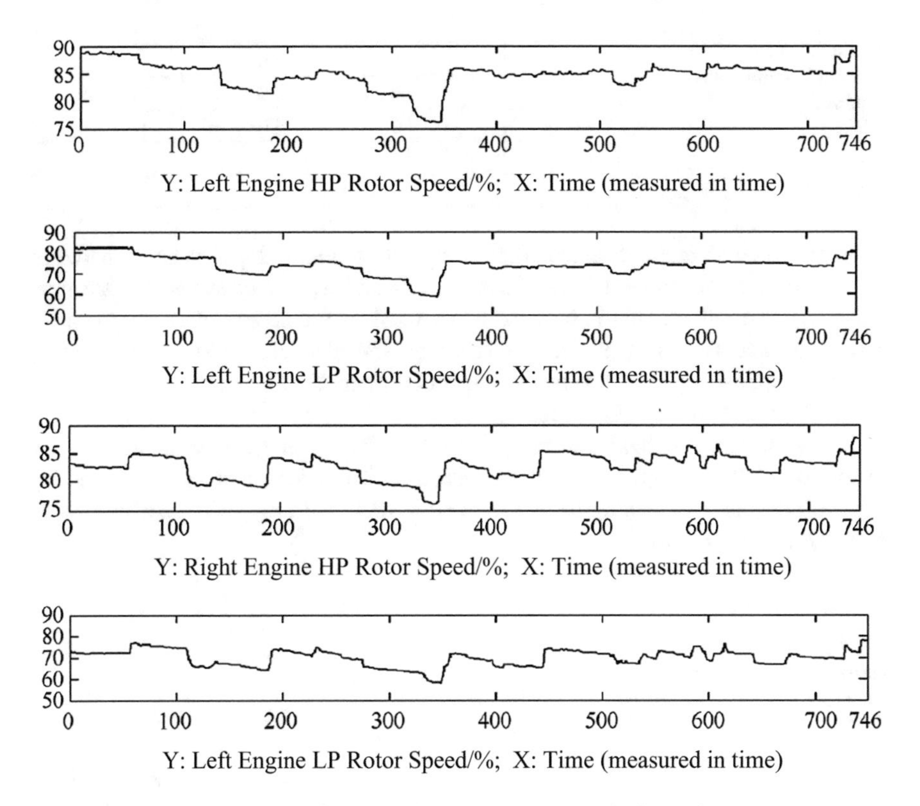

Fig. 4.24 Speed-time curves: left/right engine HP/LP rotor speed of a sortie

Fig. 4.25 Hierarchical classification based on Euclidean distance

In the first method and as shown in Fig. 4.25, a hierarchical classification of the corresponding series is obtained by calculating Euclidean distance between them. Since this method is highly data sensitive, it is incapable of distinguishing n_{1l} from n_{1r}, nor n_{2l} from n_{2r}. Even though there is only a small Euclidean distance between n_{1l} and n_{1r}, and between n_{2l} and n_{2r} as well, they should not be considered as the same type.

Table 4.4 Calculation results of original slope distance

	n_{2l}	n_{1l}	n_{2r}	n_{1r}
n_{2l}	0	0.0491	0.0859	0.1269
n_{1l}		0	0.0946	0.1419
n_{2r}			0	0.1152
n_{1r}				0

In the second method, calculation results of original slope distance between series are listed in Table 4.4, in which n_{2l}, n_{1l} and n_{2r} are classified as one type while n_{1r} as another. This is because when most of the slope values stay near to zero, plus and minus signs are neglected by the algorithm. Consequently, two curves in opposite trends turn out to be similar ones. Hierarchical classification of the method is shown in Fig. 4.26:

As to the third method and with respect to the 20 line segments, calculation results of included angle distance are listed in Table 4.5. And, as indicated by Fig. 4.27, n_{2l} and n_{1l} belong to the same type, and so do n_{2r} and n_{1r}, which agree with the real situation.

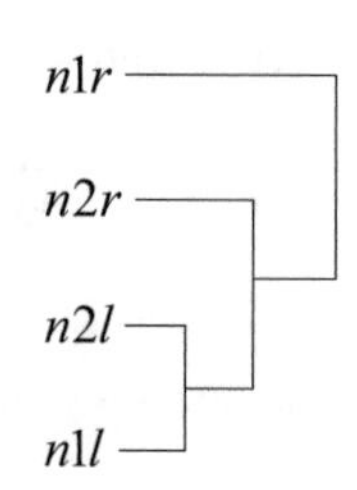

Fig. 4.26 Hierarchical classification based on slope distance

Table 4.5 Calculation results of included angle distance

	n_{2l}	n_{1l}	n_{2r}	n_{1r}
n_{2l}	0	0.262	0.350	0.488
n_{1l}		0	0.463	0.341
n_{2r}			0	0.284
n_{1r}				0

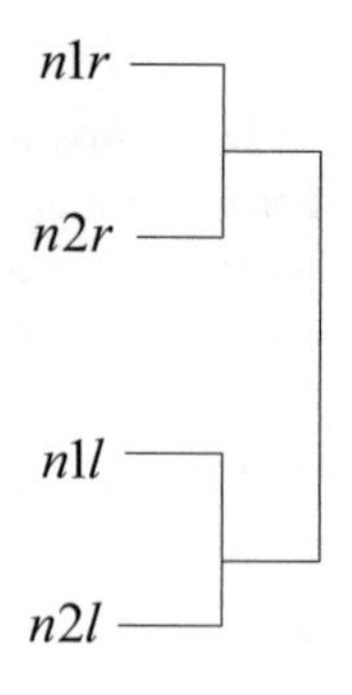

Fig. 4.27 Hierarchical classification based on included angle distance

It can be seen from the experiment results that on the basis of PLR, the use of included angle set between adjacent line segments is an innovative method of time series description. The included angle distance works very well in measuring the level of similarity of the change trends of the corresponding time series, since it overcomes the weakness of poor robustness and vague physical conception in the use of point distance. Therefore, it provides reliable technological support for pattern discovery and extraction of association rules.

4.4 Similarity Search Method in Time Series Based on Curvature Distance

Following the discussion of included angle distance in the previous section, this section will focus on the a new similarity measuring method based on curvature distance, a method that features clear physical conception, independence of values of series elements, and thus fully reflects association between continuous time points. Additionally, this method also characterized with low noise, good visualizability, as well as invariance in translation and rotation.

4.4.1 Data Preprocessing

1. Forward Translation of Time Series

In its practical application of the method, and in order for a fitting model to be established, a forward translation is necessary, that is, the original series $\boldsymbol{X} = [x_1, x_2, \ldots, x_n]$ is transformed into a nonnegative series $\boldsymbol{X}'$ when $x_i \geq 0$; $\boldsymbol{X}' = \boldsymbol{X} + |\mathbf{min}(x_i)|\boldsymbol{I}$ when $x_i < 0$. $\boldsymbol{I}$ in this expression is a unit vector, and the forward-transformed time series preserve the curve shape of the original.

2. Accumulated Generating Operation (AGO) of Time Series

After going through AGO, the nonnegative time series assume the characteristic of approximate exponential growth, which reduces randomness and is probabilistically and statistically significant. More importantly, association between transformed and original series is well preserved.

The method of 1-AGO is described as follows:

Suppose $\boldsymbol{X}^{(0)} = [\boldsymbol{X}^{(0)}(1), \boldsymbol{X}^{(0)}(2), \ldots, \boldsymbol{X}^{(0)}(n)]$ is an original nonnegative series, and apply 1-AGO to it, a new series $X^{(1)}$ is obtained:

$$\boldsymbol{X}^{(1)} = \left[\boldsymbol{X}^{(1)}(1), \boldsymbol{X}^{(1)}(2), \ldots, \boldsymbol{X}^{(1)}(n)\right] \tag{4.33}$$

In this expression, $X^{(1)}(k) = \sum_{i=1}^{k} X^{(0)}(i)$, $i = 1, 2,\ldots, n$.

Let $X^{(1)}(1) = X^{(0)}(1)$, and apply to $X^{(1)}$ IAGO (inverse accumulated generating operation), the original time series $X^{(0)}$ will be restored as follows:

$$X^{(0)}(k) = X^{(1)}(k) - X^{(1)}(k-1) \tag{4.34}$$

It can be seen from 4.33 and 4.34 that in mutual transformation between $X^{(0)}$ and $X^{(1)}$, information in the original series is losslessly preserved, and thus modeling of $X^{(1)}$ will reveal the way in which $X^{(0)}$ information changes.

3. Fitting Model of Time Series

As just mentioned above, application of 1-AGO to nonnegative series $X^{(0)}$ generates series, which assumes the characteristic of approximate exponential growth. However, considering that description of $X^{(1)}$ in terms of an exponential rule is a rather rough description, a fitting model of $X^{(1)}$ is built as follows for the purpose of improving fitting precision:

$$X^{(1)}(k) = \alpha_1 e^{\beta k} + \alpha_2 k^3 + \alpha_3 k^2 + \alpha_4 k + \alpha_5 \tag{4.35}$$

Its continuous representation can be expressed as:

$$y(t) = \alpha_1 e^{\beta t} + \alpha_2 t^3 + \alpha_3 t^2 + \alpha_4 t + \alpha_5 \tag{4.36}$$

In 4.36, α_i $(i = 1, 2, \ldots, 5)$ and β are model parameters, and $k = 1, 2,\ldots, n$.

This model is advantageous in that model parameter α_i is self-adaptive in response to different series modeling. This advantage enables fitting curve $y(t)$ to assume a mixed feature respectively of exponent, parabola, third power curve, and parabola, making possible a more precise representation of the time series.

(1) Solving Model Parameter β

For a convenient representation of the steps of solution, expression 4.35 is rewritten as follows:

$$A(k) = \alpha_1 e^{\beta k} + \alpha_2 k^3 + \alpha_3 k^2 + \alpha_4 k + \alpha_5 \tag{4.37}$$

And if

$$\left.\begin{aligned} B(k) = A(k+1) - A(k) &= \alpha_1 e^{\beta k}(e^{\beta} - 1) + \alpha_2(3k^2 + 3k + 1)k^3 + \alpha_3(2k+1) + \alpha_4 \\ C(k) = B(k+1) - B(k) &= \alpha_1 e^{\beta k}(e^{\beta} - 1)^2 + 6\alpha_2(k+1)k^3 + 2\alpha_3 \\ D(k) = C(k+1) - C(k) &= \alpha_1 e^{\beta k}(e^{\beta} - 1)^3 + 6\alpha_2 \\ E(k) = D(k+1) - D(k) &= \alpha_1 e^{\beta k}(e^{\beta} - 1)^4 \end{aligned}\right\} \tag{4.38}$$

then

$$F(k) = E(k+1)/E(k) = e^{\beta} \tag{4.39}$$

Expression 4.39 leads to:

$$\beta = \ln F(k) \tag{4.40}$$

In expression 4.40, $F(k) = \frac{A(k+5)-4A(k+4)+4A(k+3)-A(k+1)}{A(k+4)-4A(k+3)+4A(k+2)-A(k)}$, and k = 1, 2,…, n.

(2) Solving Model Parameter α_i

Use least square method to solve α_i (i = 1, 2,…, 5) and let $f(k) = e^{\beta k}$, then:

$$H = \begin{bmatrix} f(1) & 1 & 1 & 1 & 1 \\ f(2) & 2^3 & 2^2 & 2 & 1 \\ \vdots & \vdots & \vdots & \vdots & \vdots \\ f(n) & n^3 & n^2 & n & 1 \end{bmatrix}, \quad Z = \begin{bmatrix} X^{(1)}(1) \\ X^{(1)}(2) \\ \vdots \\ X^{(1)}(n) \end{bmatrix},$$

$$\hat{\alpha} = (H^{T}H)^{-1}H^{T}Z = [\alpha_1, \alpha_2, \alpha_3, \alpha_4, \alpha_5]^{T} \tag{4.41}$$

4.4.2 Representation of Curvature Set for Time Series

1. Piecewise Six-Dimension Representation (PSR) of Time Series

PLR is normally used in feature extraction and dimension reduction of time series. PLR, proposed by Keogh and others, is a from bottom to up algorithm that has solved the problem the number of segments a time series should be segmented into. On the basis of PLR, Piecewise Six-dimension Representation (PSR) will be proposed in the present section, which works in a way described below:

(1) Piecewise Dimension reduction

Apply PLR to the original time series $X^{(0)}$:

$$X^{(0)} = \left|(X_1^{(0)}, t_2), (X_2^{(0)}, t_3), \ldots, (X_i^{(0)}, t_{i+1}), \ldots, (X_{n-1}^{(0)}, t_n)\right| \tag{4.42}$$

In this expression, t_i is the time when the ith segment of time series stops; n is the number of the segments; and $X_i^{(1)}$ $(i = 1, 2, \ldots, n-1)$ is the ith segment with $\mathbf{dim}(X_i^{(1)}) \geq 6$.

(2) Dimension Reduction of Each Segment

Four data points, excluding endpoints of each segment, will be selected from each segment $X_i^{(0)}$ $(i = 1, 2,\ldots, n - 1)$ to generate a new piecewise series $\widetilde{X}_i^{(1)}$. PSR of a time series is often expressed as follows:

$$X^{(0)} = \left|(\widetilde{X}_1^{(0)}, t_2), (\widetilde{X}_2^{(0)}, t_3), \ldots, (\widetilde{X}_i^{(0)}, t_{i+1}), \ldots, (\widetilde{X}_{n-1}^{(0)}, t_n)\right| \tag{4.43}$$

In this expression, t_i is the time when the ith segment of time series stops; $(n - 1)$ is the number of the segments; and $\widetilde{X}_i^{(1)}$ $(i = 1, 2, \ldots, n - 1)$ is the ith PSR segment with $\mathbf{dim}(\widetilde{X}_i^{(1)}) = 6$.

2. Curvature Set of Time Series

Definition 4.8 For PSR expression $X^{(0)} = \left|(\widetilde{X}_1^{(0)}, t_2), (\widetilde{X}_2^{(0)}, t_3), \ldots, (\widetilde{X}_i^{(0)}, t_{i+1}),\right.$ $\left.\ldots, (\widetilde{X}_{n-1}^{(0)}, t_n)\right|$, application of 1-AGO to $\widetilde{X}_i^{(0)}$ generates $\widetilde{X}_i^{(1)}$, which is then fitted with (4.36). Thus, an approximate expression of a curvature series can be obtained by fitting each time point of $\widetilde{X}_i^{(1)}$ with curvature of the curve. This approximate expression can be defined as $X^{(1)} = |\tilde{c}_1, \tilde{c}_2, \ldots, \tilde{c}_{n-1}|$, in which $\tilde{c}_i = \{c_t\}$, $\mathbf{dim}(\tilde{c}_i^{(1)}) = 6$, and $c_t = \frac{\tilde{y}_i^{(1)\prime\prime}}{(1+\tilde{y}_i^{(1)\prime})^{3/2}}$ is the curvature of in-segment time points of $\tilde{y}_i^{(1)}$ which is a nonlinear model curve corresponding to $\widetilde{X}_i^{(1)}$. And in this expression $\tilde{y}_i^{(1)} = \tilde{\alpha}_1 e^{\tilde{\beta}t} + \tilde{\alpha}_2 t^3 + \tilde{\alpha}_3 t^2 + \tilde{\alpha}_4 t + \tilde{\alpha}_5$, $\tilde{\alpha}_j (j = 1, 2, \ldots, 5)$ and $\tilde{\beta}$ are parameters of the nonlinear model curve corresponding to $\widetilde{X}_i^{(1)}$.

Transformation of PSR of $X^{(1)}$ into a curvature set representation realizes $R^N \to R^{6(n-1)}$ mapping, in which N is the number of dimensions of $X^{(1)}$, and normally $N \geq 6(n - 1)$. The mapping from $\mathbf{R}^N$ to $\mathbf{R}^{6(n-1)}$ improves efficiency of dimension reduction.

3. Dimension Time Alignment of Curvature Set

Normally, effective similarity search can be done only when two time series are aligned in both dimension and time.

Definition 4.9 Dimension time alignment means that in PLR of the two time series to be compared, corresponding segments should match in number of dimensions, as well as starting and ending time. To be more specific, time series data extracted from paired segments for comparison should be aligned in time.

As illustrated by Fig. 4.28, corresponding segments X_i in time series A and B have the same starting and ending time, as well as the same number of dimensions, 12. Dimension time alignment of X_i results in $\widetilde{X}_i$, in which 6 data are time aligned at all time points within the segments in comparison.

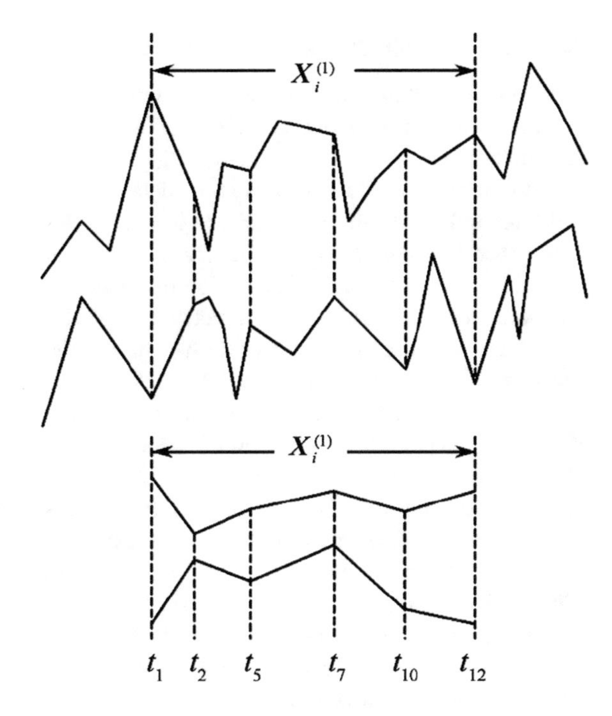

Fig. 4.28 Dimension–time alignment in time series *A* and *B*

4.4.3 Curvature Distance Measurement for Time Series

1. Curvature Distance

Curvature is the rotation rate of a tangential direction angle to radian length at a curve point, indicating the rate of curving at the point. In a similar way, the curvature at a time point of a fitting curve of time series accurately reveals tendency of the time series in the time neighborhood. Therefore, curvature distance does well in representing similarity in time series: the closer the curvature distance, the higher the similarity will be. Curvature distance can be defined as follows:

Definition 4.10 If $X = |\tilde{c}_1, \tilde{c}_2, \ldots, \tilde{c}_{n-1}|$ and $X' = |\tilde{c}'_1, \tilde{c}'_2, \ldots, \tilde{c}'_{n-1}|$ are two time-aligned time series of equal length that are represented in form of two curvature sets, the curvature distance $D_c(X, X')$ between them is:

$$D_c(X, X') = \frac{\sum_i^n |\tilde{c}_i - \tilde{c}'_i|}{n-1} \tag{4.44}$$

Basically, curvature distance should meet the following requirements:

(1) Nonnegativity: $D_c(X, X') > 0$, $X \neq X'$;
(2) Self-similarity: $D_c(X, X') = 0$, $X = X'$;
(3) Symmetry: $D_c(X, X') = D_c(X', X)$;
(4) Triangle inequality: $D_c(X, X') \leq D_c(X, X'') + D_c(X', X'')$.

2. Noise Reduction of Curvature Distance

Curvature distance is insensitive to noise, and to a certain degree capable of noise reducing. More importantly, it remains stable and effective in the case of noise interference.

Noise reduction of time: Due to time delay in sampling, the actual sampling time and time of data recording may not be identical. However, since curvature distance only concerns itself with the tendency of a small neighborhood of a curve point, only time noise at the point in question may cause change to curvature distance. At the same time, because time length taken by the selected points is much shorter than that of the whole series, time noise has very little effect on the measurement of curvature distance, and curvature distance works very well in reducing time noise

Piecewise noise reduction: PSR is a time series representation method based on PLR, which is good at data compression and noise filtering. By means of managing the amount of segments to be divided, PLR functions in such a way that it filters noise where necessary. For low level noise, more segments can be selected for a more detailed representation of the time series; for high level noise, fewer segments should be selected to improve the estimated stability of fitting model parameters. Thus, PLR functions like a low pass filter when time noise is at a high level.

Principle of noise reduction: Curvature distance is calculated on the basis of the curvature of the fitting curves of time series. Since curvature itself is a differential coefficient of the expression of a fitting curve, it may serves as a noise filter in principle.

3. Algorithm of Similarity Search

In time series research, similarity search is a search in database for series that are similar in pattern with the given series. Usually a preset similarity threshold is required for the search to be done. As to the present algorithm, similarity search targets a given time series for any time series, curvature distance of which satisfies the condition of $D_c(\boldsymbol{X}, \boldsymbol{X}') \leq \varepsilon$. The algorithm is described as follows:

Step 1: According to the method of forward translation proposed in Sect. 4.4.1, apply to forward translation to both $\boldsymbol{X}$ and $\boldsymbol{X}'$.

Step 2: According to PSR proposed in Sect. 4.4.2, apply PSR to both $\boldsymbol{X}$ and $\boldsymbol{X}'$.

Step 3: According to 1-AGO in Sect. 4.4.1, apply 1-AGO to both $\boldsymbol{X}$ and $\boldsymbol{X}'$.

Step 4: According to Definition 4.8, represent $\boldsymbol{X}$ and $\boldsymbol{X}'$ in terms of curvature set.

Step 5: According Definition 4.9, apply dimension time alignment to curvature sets of $\boldsymbol{X}$ and $\boldsymbol{X}'$.

Step 6: According to 4.10, calculate $\boldsymbol{D}_c(\boldsymbol{X}, \boldsymbol{X}')$; search all the time series within the threshold ε, and group the search results into a set.

Similarity threshold ε has a big influence on search results, especially on the search based on the index. In theory, threshold ε may be any positive numbers; but in practice, its choice is rather complicated and its determination is subject to the specific application background, disciplinary knowledge involved, and expertise.

4.4.4 Verification on Clustering of Flight Data Based on Curvature

For verification of the algorithm on the left engine of a certain aircraft, four groups of time series data of exhaust temperature are used. 2000 data are extracted from each data group to obtain four subseries, that is, X_1–X_4. For a better intuitive comparability of the data, normalization processing, which exerts no influence on physical characteristics of the flight data, is applied to the four time series, positioning each between 0 and 1. Curves of each time series are shown in Fig. 4.29.

Experiment 1: In order for efficiency and precision of the nonlinear model to be verified, the first 9 data of time series X_1 are chosen for modeling. Following the building of a fitting model of the series is the solution and fitting of model parameters. By solving the model, a fitting curve equation is derived as follows:

$$y(t) = 1.2216e^{-0.3194t} + 0.0013t^3 - 0.0106t^2 + 0.6399t - 1.2188$$

As shown in Table 4.6, it can be seen from a comparison between model calculation results and data series that nonlinear model is advantageous for its small error and high fitting precision. At the same time, it is observed that while expression (4.37) is used in five iterative computations for solving model parameter β, only six data are needed for building a fitting model. For sake of modeling precision, 6 dimensions are chosen for piecewise representation; this is how PSR gets its name.

Experiment 2: Euclidean distance is the most widely used method for measuring point-to-point distance, while pattern distance is typically used in current research of similarity measurement of time series patterns. In order to verify the independence of curvature distance on series elements and the efficiency of pattern measurement, comparisons are made through experiments among four methods of distance measuring, that is, curvature distance, two classical methods, and radian distance with high robustness and stability. In the experiments, non-piecewise Euclidean distance is calculated for comparison with four types of piecewise distance, the number of segments for which is 40. All the calculation results are listed in Table 4.7. It can be concluded that pattern distance works on the basis of three patterns of the corresponding series elements, including "rising, holding, and falling." An examination of 1000–2000 data reveals that all the X_2 elements are greater than their counterparts in X_4, which means that X_2 elements are in a "rising" mode, and that it is not appropriate to make a comparison between X_2 and X_4. Even though the calculation results show that X_1 and X_2 are most similar to each other, this result does not agree with that of manual interpretation. Meanwhile, calculation results of curvature distance, radian distance, and non-piecewise Euclidean distance jointly conclude that X_1 and X_3 are most similar to each other. Due to the dependence of Euclidean distance on the values of series elements, piecewise Euclidean distance may be interpreted as different from its non-piecewise counterpart, which

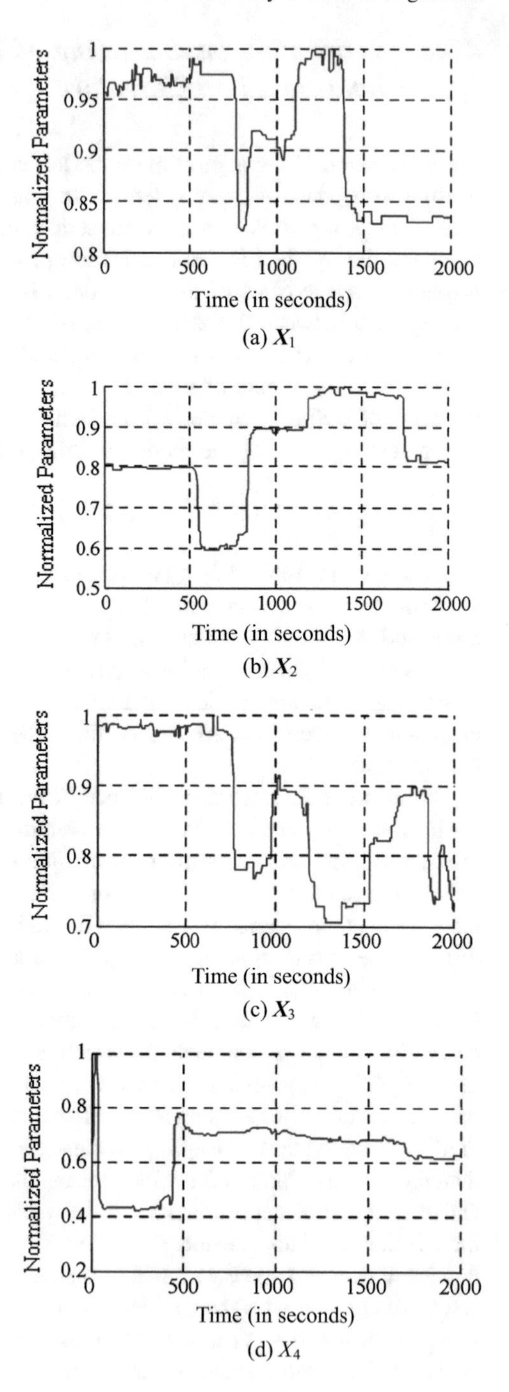

Fig. 4.29 Four groups of flight data as time series

Table 4.6 1-AGO data series and model calculation results series

Sequence no.	1	2	3	4	5	6	7	8	9
Data series	0.3000	0.3709	0.4392	0.4845	0.5264	0.5705	0.6158	0.6622	0.7062
Model series	0.3000	0.3735	0.4348	0.4847	0.5290	0.5713	0.6143	0.6598	0.7093
Model series	0	0.0026	−0.0044	0.0002	0.0026	0.0008	−0.0015	0.0076	0.0031

Table 4.7 Comparison of Euclidean, pattern, radian, and curvature distance

Distance type	$D(X_1, X_2)$	$D(X_1, X_3)$	$D(X_1, X_4)$	$D(X_2, X_3)$	$D(X_2, X_4)$	$D(X_3, X_4)$
Euclidean D_E (non-piecewise)	0.2998	0.0659	0.6019	0.2955	0.5288	0.5998
Euclidean D_E (30 segments)	1.2108	1.768	2.4441	2.3000	2.0702	3.9165
Mode D_M (30 segments)	0.2833	0.9833	1.0000	0.9500	1.0000	1.0000
Radian D_R (30 segments)	0.2833	0.1459	0.1654	0.2314	0.2138	0.1558
Curvature D_C (30 segments)	0.1066	0.0537	0.1329	0.1015	0.1856	0.1421

highlights the independence of curvature distance and radian distance on series elements.

Experiment 3: In order to verify the measuring stability and precision of curvature distance in cases where there are different piecewise representations, apply PLR respectively to the four data series in the sequence of 15, 30, 50, and 75 segments, and calculate five types of distance. The calculation results are listed in Table 4.8, analysis of which comes to the following conclusions: calculations of four types of distance (Euclidean, Curvature, Slope, and Radian) for different segmentations indicate that X_1 and X_3 are most similar to each other; as for included angle distance, it reaches the same interpretation results as the former three distance types in cases where there are 30, 50, and 75 segments, verifying the stability of interpretation results of curvature distance; calculation results of curvature distance for different segmentations reveal that X_3 and X_4 are most different from each other, so do calculation results of slope and radian distance at high resolutions, as well as manual interpretations, verifying the precision of curvature distance in identifying time series which are most different from each other. It can be seen from the above conclusions that compared with other types of distance, curvature distance features stability and precision in similarity interpretation of time series.

Experiment 4: This experiment is designed for a quantitative research of the five methods of distance measurement with respect to their advantages and disadvantages so that stability of the five methods can be evaluated in terms of convergence in response to changing length of data. It can be seen from Table 4.8 that in distance measuring between paired data series, four methods, that is, Euclidean, slope, included angle and radian, display a property of divergence as the number of

Table 4.8 Comparison of 5 types of distance for different segmentations

Segmentation	Distance type	$D(X_1, X_3)$	$D(X_1, X_4)$	$D(X_2, X_3)$	$D(X_3, X_4)$
15	Euclidean D_E	1.2854	1.7699	1.7102	1.4586
	Slope D_S	0.4754	0.4871	0.8610	0.6345
	Included angle D_a	0.1313	0.1168	0.1777	0.1534
	Radian D_R	0.1046	0.1388	0.221	0.1493
	Curvature D_C	0.0758	0.1331	0.1806	0.1854
30	Euclidean D_E	1.768	2.4441	2.3000	2.0702
	Slope D_S	0.6352	0.7700	1.0205	0.9586
	Included angle D_a	0.2502	0.2750	0.7152	0.6774
	Radian D_R	0.1459	0.1654	0.2314	0.2138
	curvature D_C	0.0537	0.1329	0.1015	0.1856
50	Euclidean D_E	2.2943	3.1188	2.8931	2.6921
	Slope D_S	0.8900	0.9367	1.2068	1.2323
	Included angle D_a	0.8131	0.9263	1.5357	1.5165
	Radian D_R	0.1837	0.1838	0.2513	0.2589
	Curvature D_C	0.0918	0.1142	0.1116	0.1290
75	Euclidean D_E	2.7900	3.8016	3.5942	3.1928
	Slope D_S	0.9946	1.2814	1.5508	1.7362
	Included angle D_a	1.8151	2.0648	4.0310	3.6252
	Radian D_R	0.2148	0.2404	0.3603	0.3838
	Curvature D_C	0.0852	0.1459	0.1088	0.1560

segments increases. That is to say, unlike the other four methods, curvature distance does not display a tendency of divergence in its calculation results as resolution increases. A comparison of the five types of distance shows that while curvature distance displays a property of good stability without divergence, Euclidean and included angle distance are characterized with fast divergence; and that slope and radian distance are slow in divergence and poor in stability.

For a further research of the property of convergence of curvature distance in response to change of data length, X_1 and X_3 are chosen for verification which is done respectively for relative and absolute change of data length, and in which different statistics are used as measures.

(1) Relative variation of data length: On the basis of the briefly indentified similarity between two time series with unchanged data length, a more precise account of the relationship of similarity can be obtained by increasing the number of segments because with the increasing of resolution, the total data length for calculation increases as a result. In this situation, standard deviation can be used to measure the convergence of curvature distance.

Let the length of X_1 and X_3 be 2000, and calculate curvature distance corresponding to different segments. Calculation results are listed in Table 4.9.

Table 4.9 Curvature distance D_c (X_1, X_3) for different number of segments

Segments	5	10	15	25	30	45	50	75	90
$\boldsymbol{D_c(X_1, X_3)}$	0.1071	0.1207	0.0759	0.1850	0.0538	0.1487	0.0919	0.0853	0.0869

As indicated by Table 4.9, when segments range from 5 to 45, curvature distance is in a bigger range of variation because of low resolution; when segments range from 50 to 90, curvature distance is in very small range of variation as a result of high resolution. These calculation results not only justify curvature distance as an ideal method of distance measurement, but also reveal high stability of curvature distance in response to relative change of data length.

(2) Absolute variation of data length: When the number of segments remains the same, that is, the resolution is unchanged, the length increase of time series means absolute increase of data length, and consequently the relationship of similarity between time series may fall into a state of uncertainty caused by the involvement of new data. In this situation, it is appropriate to use harmonic mean value for description of convergence of curvature distance in state of uncertainty. Let the number of segments constantly be 12, and calculate curvature distance between $\boldsymbol{X}_1$ and $\boldsymbol{X}_3$ when there is a continuous increase of series length. The calculation results are listed in Table 4.10.

As indicated by Table 4.10, when the resolution remains the same and data length is 1200, the smallest curvature distance occurs, and thus $\boldsymbol{X}_1$ and $\boldsymbol{X}_3$ are most similar to each other; when data length ranges from 1350 to 1800, there appears greater curvature distance, and similarity between $\boldsymbol{X}_1$ and $\boldsymbol{X}_3$ decreases. Table 4.10 also indicates that when data length extends into the range beyond 1800, similarity between $\boldsymbol{X}_1$ and $\boldsymbol{X}_3$ increases together with the increase of data length, which agrees with the real situation. In summary, the increase of data length may lead to uncertainty of similarity between time series.

In calculating data length variation, harmonic mean value of curvature distance changes in a way as shown in Fig. 4.30. When data series grow in length, harmonic mean value of curvature distance fluctuates in a very small range of 0.2332–0.2612. As indicated by Fig. 4.30, the method of curvature distance displays a certain property of convergence when there is absolute variation in data length.

By taking into full consideration the information of series shape, curvature distance features not only high efficiency and precision, but also independence on series elements, together with good ability of noise reduction and robustness. As indicated by the statistics of measurement results at different resolutions, curvature distance displays good convergence and stability, and can be put into effective use in such areas as similarity query, mode discovery, and mode clustering.

Table 4.10 Curvature distance D_c (X_1, X_3) for different series lengths

Series length	150	300	450	600	900	1050	1200	1350	1500	1650	1800	1950
D_c (X_1, X_3)	0.237	0.234	0.273	0.218	0.229	0.269	0.193	0.322	0.334	0.314	0.354	0.234

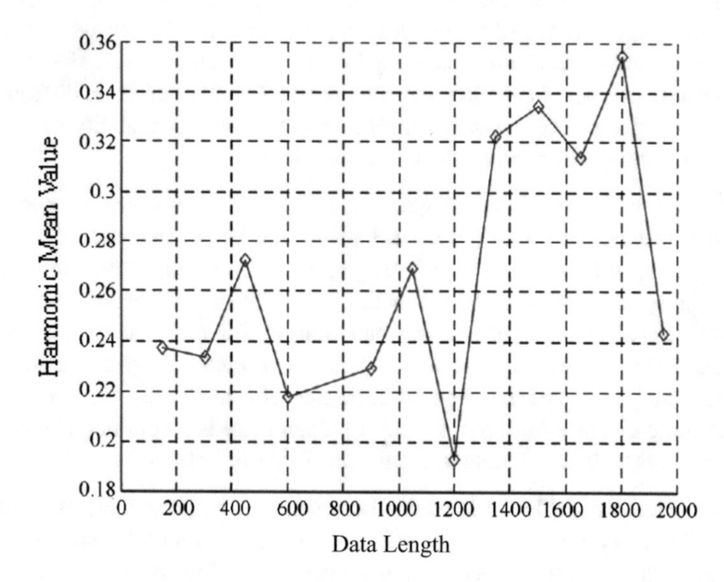

Fig. 4.30 Variation of harmonic mean value of curvature distance in response to data length change

4.5 Multivariable Flight Data Similarity Search Method Including Changing Step to Set Data in Different Bins

Owning to its quick extraction of segment characteristics, the method of piecewise linear representation has attracted much attention in the field of flight data research. In the present section and on the basis of this method, a multivariable similarity search method is proposed for data in different bins, which is a WSTB (Weighted Shape To Bit-vector) and an extension of piecewise linear representation into multivariable similarity search.

4.5.1 Piecewise Linear Representation for Time Series

Suppose that time series $\boldsymbol{A}$ is obtained by doing k point sampling to a time series, $\boldsymbol{A}'$ is the piecewise linear representation of $\boldsymbol{A}$, and $\boldsymbol{A}' = \{\boldsymbol{A}_{x1}, \boldsymbol{A}y_1, \boldsymbol{A}x_r, \boldsymbol{A}y_r\}$. Random segment of time series $\boldsymbol{A}$ is expressed as $(\boldsymbol{A}_{x1}, \boldsymbol{A}x_r)$ and $(\boldsymbol{A}y_1, \boldsymbol{A}y_r)$. An

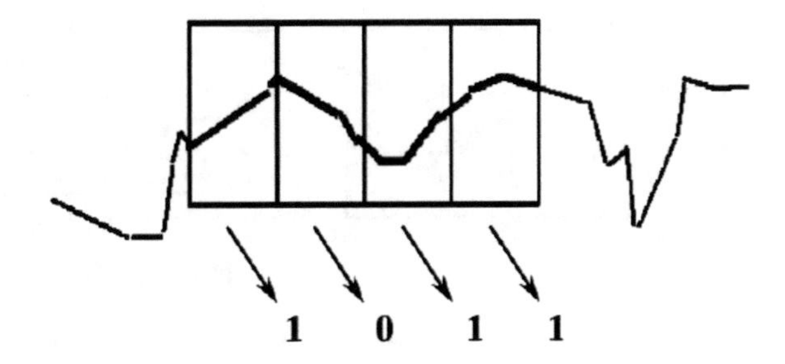

Fig. 4.31 Windows moving along time series

index built on the basis of piecewise linear representation is illustrated in Fig. 4.31. While moving windows with fixed length along the original series, and examination of the series parts that appear in the windows results in a character string (1011 in Fig. 4.31). This character string can be seen as the main identification of the series in question. Each data segment is divided into two parts, that is, "up" and "not-up" represented, respectively, by 0 and 1. In this way, query result can be obtained directly.

However, in adoption of the method above, the following problems concerning representation of multivariable time series remain unsolved:

(1) With the increase of data size, linear representation of the curves of multivariable time series will give rise to extremely big amount of data groups, management of which requires special way of data saving and representation.
(2) For similarity search of multivariable time series different in length but similar in trend, repeated index building is time consuming and therefore, inefficient.

For the reasons above and on the basis of linear representation, a binning storage method is proposed, by means of which time series data are stored in different bins according the index. Inside the bins time series data are automatically sorted according to the distance so that problem of data storage and fast indexing can be solved.

4.5.2 *Indexing Tag Based on Changing Step to Set Data in Different Bins*

To ensure the expansibility of the algorithm, adjustment should be made to the search method. The emphasis of adjustment is on the use of original data index table for changing step size. Since a data index table should be built for every similarity search of unary data, unary data similarity search is not applicable to multivariable and high-dimensional data. Moreover, no systematic study on window length has been done in the widely recognized algorithm of unary data similarity search, which limits the application of the algorithm in a narrow area.

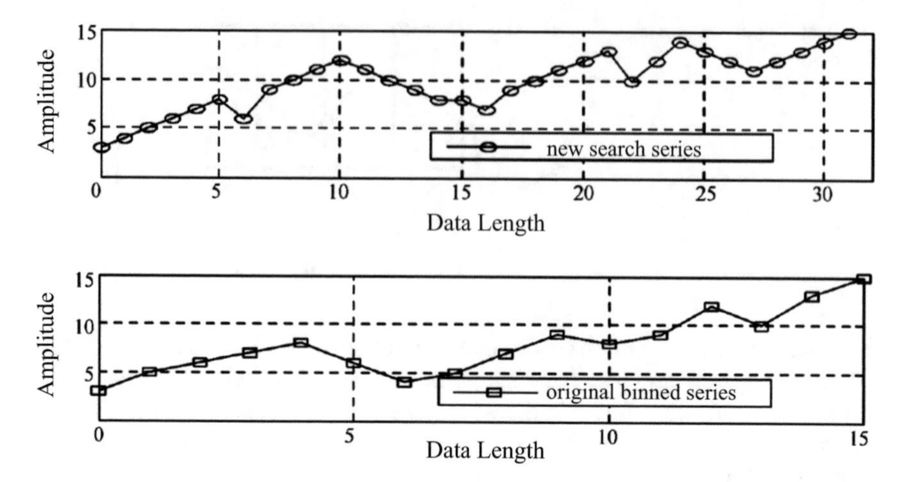

Fig. 4.32 Changing step search

On the basis of the method of binning index, a new method is here proposed to combine changing step with original data index. Figure 4.32 is an illustration of indexing of changing step search, in which the horizontal ordinate is for length, and the vertical ordinate for amplitude. It should be noted that the original index of 1011 is built on a curve with 16 data points while the new series to be searched is made of 32 data points. In the new search, slide the windows as it had been done with the original series and set the grid at 8, a tag of 1011 is also obtained. However, due to different search length and amplitude, different weight coefficients should be used to ensure the effectiveness of the search.

The following points are worthy of notice in actual search:

(1) Supplementary condition should be specified for changing step search. It may happen in changing step search that the length of sample data is not divisible by the chosen step size. In the original series, data length is l, number of sampling points is i, and length of indexing is b; therefore, when sampling is done for every j sampling points of the new search series, the index should be supplemented on the condition stated below:

$$n = l \times i \times b \tag{4.45}$$

(2) Tail indexing tags should be added to ensure equal length of indexing. When the supplemented number of sampling points is smaller than i, the tai indexing label should be 0 or 1. In Fig. 4.32, choose the first 30 data for indexing labels; if step size is changed to 8, the indexing labels will be unobtainable. Adding to the data tail indexing labels results in two different indexing results, 1011 and 1010.

(3) Method of choosing optimal step size. In order to prevent multiple indexes from occurring, the following way is used to determine the step size:

$$D_s = \begin{cases} D_0/k & \text{if } \operatorname{Mod}(L/D_0) < 0.5D_0 \\ D_0 & \text{else} \end{cases} \tag{4.46}$$

In this expression, $\boldsymbol{D}_0$ is the current step size, $\boldsymbol{D}_s$ the actual indexing step size, and $\boldsymbol{L}$ the actual length.

4.5.3 Similarity Search Method Including Changing Step to Set Data in Different Bins

On the basis of analyzing the indexed database and index sampling, this algorithm specifies ways of data indexing under different conditions. Following the procedures outlined below, data curve positioning is done by adopting the method of index matching and in relation to the step size of different windows.

(1) Select the data segments concerned and build index sampling in way of piecewise linear representation.
(2) Store the data by binning and enable self-sequencing of data curves in the bin.
(3) Determine whether the present indexing method is full sequence matching or subseries match.
(4) Obtain preliminary search results on the basis of a given distance measurement.
(5) Improve search efficiency to the maximum by using step-changing coefficient and original index tags (Fig. 4.33).

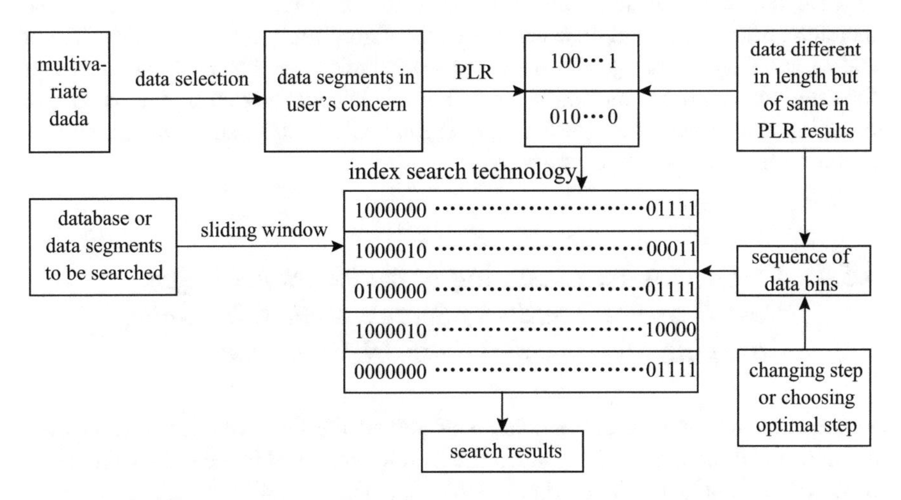

Fig. 4.33 Illustration of similarity search method including changing step to set data in different bins

In this algorithm, bin_labels stands for the index table. The following are the pseudocodes of the program part of the algorithm:

```
procedure search(bin_labels)

initialize best-so-far to infinity
get query shape Q from user
let R be the discretization vector obtained from Q
place bin_labels onto heap using min_dis(R,B) as heap key
do

best-so-far = DS/(j/i);
(N-1) bin_labels = 0 or 1

while j = 1*i
do

remove B from top of the heap
retrieve corresponding bin
in-bin-search(Q,bin,best-so-far)

while best-so-far > min_dis(R,B)

end
```

As shown in the program, the optimal values of distance appear in company with different index results in the search process, which is a diminishing one, and in which a preset initial value is not necessary for the program to run.

It can be seen that in its actual use the algorithm, index results of sampling can be incorporated into the program as a variable array, the variability of which satisfies the requirement of different index results for the same sampling, and therefore, highlights good expansibility of the algorithm. Moreover, the method of adding auxiliary tail tag greatly improves search efficiency. Establishing index tags for data bins is a time consuming work, but this is compensated by the method of changing step which makes use of the original search tags and ensures high precision by means of dynamic adjustment.

4.5.4 Verification on Clustering of Multivariable Flight Data Based on Similarity Search Method Including Changing Step to Set Data in Different Bins

Figure 4.34 is a graphic representation of part of the (technologically processed) attitude data of an aircraft from takeoff to landing, arranged in the order of angle of bank (Fig. 4.34a), heading (Fig. 4.34b), altitude (Fig. 4.34c), angle of pitch (Fig. 4.34d), and velocity (Fig. 4.34e). The horizontal axis indicates relative time.

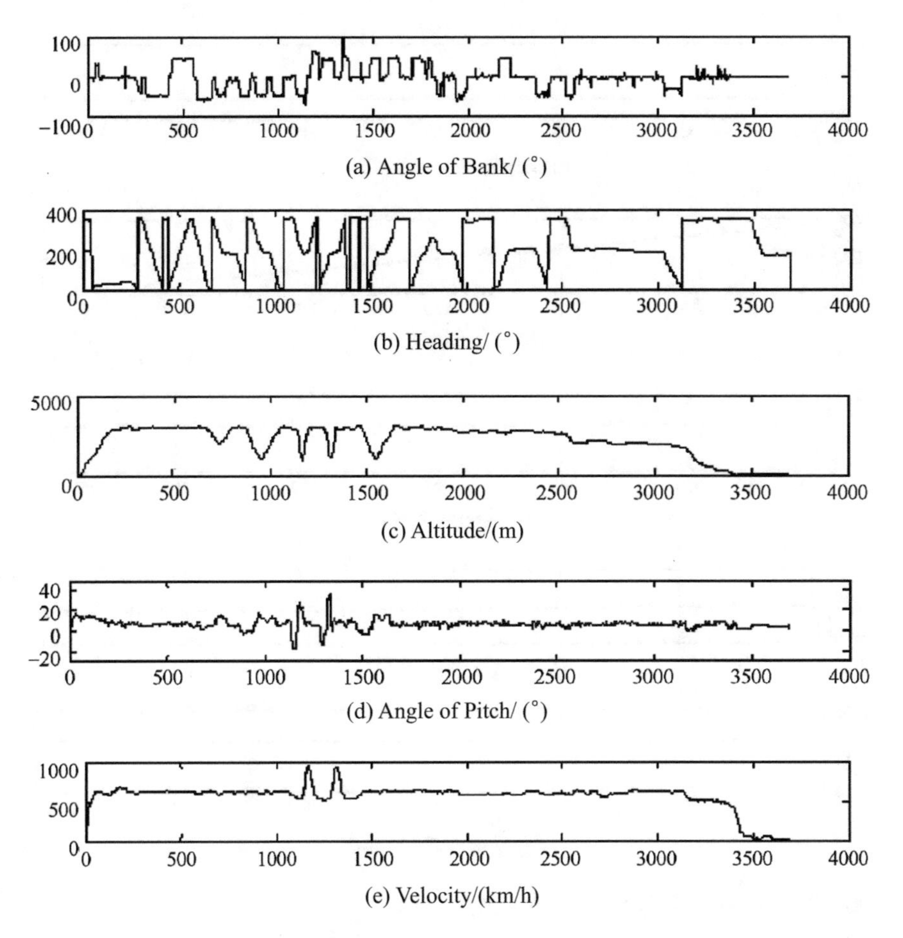

Fig. 4.34 Part of aircraft attitude data

As indicated by Fig. 4.34, aircraft maneuver is a very complicated phenomenon, including basic maneuver and tactical maneuver. Basic maneuver is further divided into two types, aerobatic flight and instrument flight. Take aerobatic flight for example, it can be classified into three major categories of maneuver: horizontal maneuver, vertical maneuver, and combination of the two. Among horizontal maneuvers, circling at maximum angular velocity, augmented circling, and horizontal snap/slow roll are typical, while half roll and half loop, immelmann turn, and half loop are typical vertical maneuvers. As to combination of horizontal and vertical maneuvers, climbing/descending roll and diving–climbing–turning at different dive angles are typical among others. The discussion that follows is aimed to see whether a precise linear representation of aircraft maneuvers is possible or not. As shown in Fig. 4.35, a set of standard flight data is used as sample data, between the two dotted lines is the space where circling at a bank angle of 60° occurs.

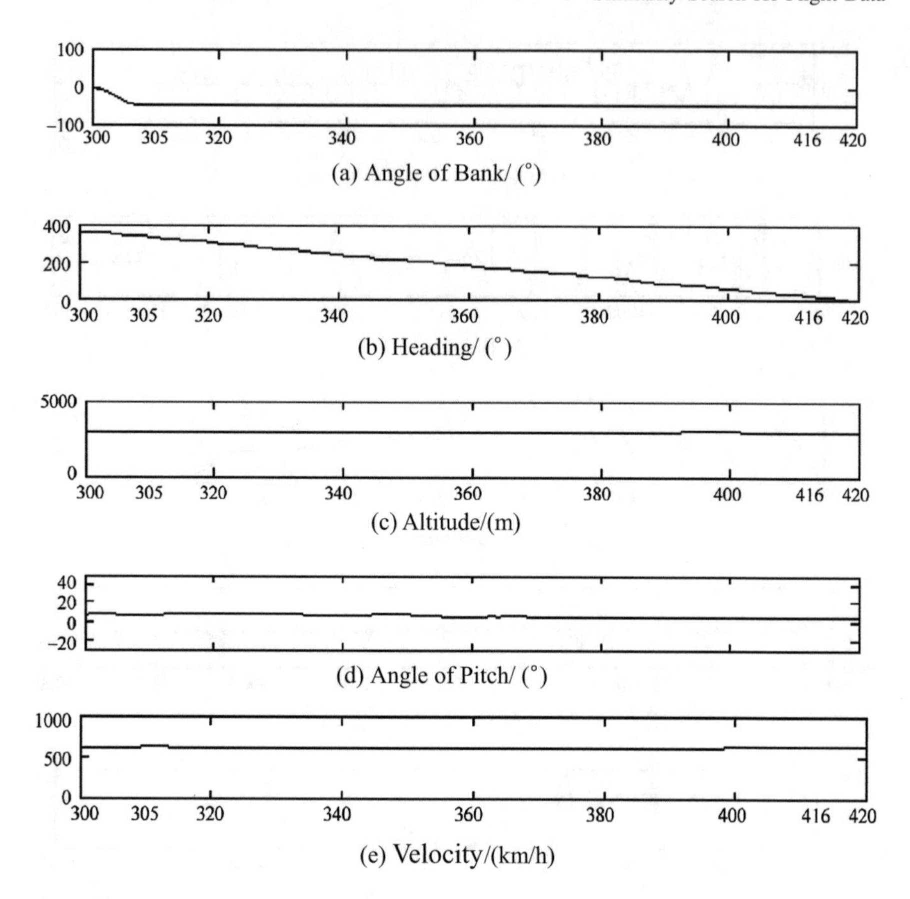

Fig. 4.35 Parameter curve of 60° bank angle circling

The flight data in Fig. 4.35 have the following properties: ① due to the complicated model of data formation, there is no intuitive association between parameters; ② in view of complexity of aircraft maneuvering and operations of equipments involved, analytical comparisons should be done between multiple models of data; ③ there is no unified boundary mark between stability and fluctuation of data; ④ due to the use of different data samples, similarity search done to a database may be unnecessarily repeated. These properties call for a search method that distinguishes itself from others.

1. Step Size Selection in Piecewise Linear Representation of Flight Data

As shown in Table 4.11, normalized representation is applied to the flight data of 60° bank angle turn, which is divided into 12 segments with an interval of 10. There are 24 segments with an interval of 5 in Table 4.12. It can be seen from Table 4.11 that in the stage of entry and diversion of a maneuvering, altitude and velocity change in opposite trends, which fully agrees with the real situation of continuous

Table 4.11 Piecewise linear representation with the interval of 10

PLR segments	Bank angle	Heading	Altitude	Pitch angle	Velocity
1	0	0	0	0	1
2	1	0	1	1	0
3	1	0	1	0	0
4	0	0	0	0	1
5	1	0	0	1	0
6	0	0	1	0	1
7	0	0	0	0	0
8	1	0	1	0	0
9	0	0	1	0	0
10	1	0	1	0	1
11	0	0	0	1	1
12	0	0	0	0	1

Table 4.12 Piecewise linear representation with the interval of 5

PLR segments	Bank angle	Heading	Altitude	Pitch angle	Velocity
1	0	0	0	0	0
2	0	0	0	0	1
3	1	0	0	1	0
4	0	0	1	0	0
5	0	0	1	0	0
6	1	0	1	1	0
7	0	0	0	0	1
8	0	0	0	0	1
9	0	0	0	1	0
10	1	0	0	0	0
11	0	0	1	0	1
12	1	0	0	0	1
13	0	0	0	1	0
14	0	0	0	0	0
15	1	0	1	1	0
16	0	0	0	0	0
17	1	0	0	1	0
18	0	0	1	0	0
19	1	0	1	0	1
20	0	0	0	0	1
21	0	0	0	1	1
22	0	0	0	0	0
23	0	0	0	0	1
24	0	0	0	0	1

turn. As to the data listed in Table 4.12, change trends of some individual parameters are not in full agreement with way of circling maneuver.

As indicated by Tables 4.11 and 4.12, in theory, the results of the same type of maneuver should be highly identical; however, because of the differences in aircraft performance, operation skills of pilots and weather conditions at the time of maneuvering, complex and trivial variations can be identified in the flight data. Moreover, data analysis reveals that for each type of maneuvering there is a corresponding data mode. In order to ensure the typicality of search samples and for the algorithm to agree with the rule of human cognition, for different flight data samples, there should be appropriate step size for PLR.

2. Similarity Search Method Including Changing Step to Set Data in Different Bins

Figure 4.36 is about an improved algorithm of flight data search method, in which comparisons are made between the first parameter index tags after the initial samples are ranked in the order of importance according to the maneuver.

Adjustment of algorithm is motivated by the following practical needs:

(1) Since flight data diverse in types with different value ranges, physical meanings, and sampling frequencies, it does not make sense to make comparisons between different types of flight data. What is significant is to make comparisons between time series formed by flight data of the same type.
(2) Data analysts usually pay more attention to flight data of certain types, which means an automatic sequencing of the data in the order of importance. For example, in searching flight data of turn maneuver, what comes into analysts' attention first is the change trend of bank angle, and the analysis is focused on whether the maneuver follows the procedure of "banking-holding-leveling off." With the help of cumulative change of heading in maneuvering, the whole process of maneuvering can be observed and examined.
(3) Variation in step size may great difference in linear representation of flight data of the same type. In order to make sure that results of flight data search satisfy

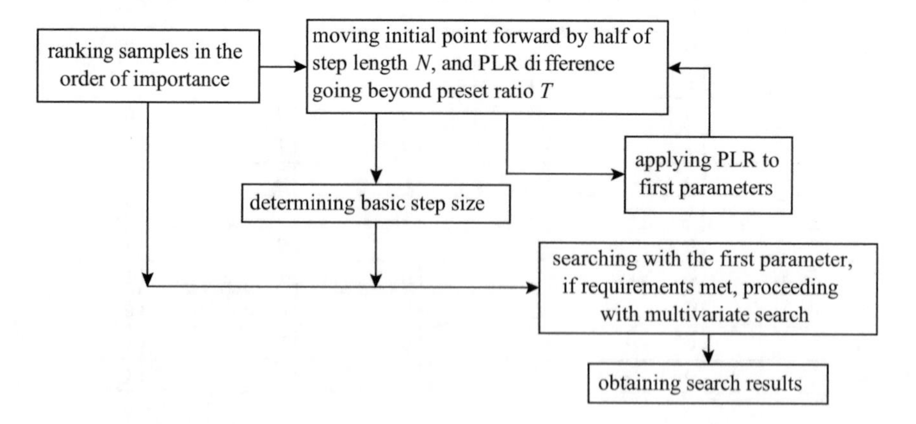

Fig. 4.36 Algorithm improvement for flight data search

practical needs, step size should be adjusted in such a way that the initial point is moved forward by half of $\boldsymbol{N}$; and when difference in PLR goes beyond the given proportion of $\boldsymbol{T}$, initial point should be moved and positioned for another time.

3. Results of Verification

Coding with Visual C++ 6.0, and visualize the data by invoking Engine Interface of Matlab 7.0; verify the efficiency and precision of WSTB (Weighted Shape To Bit-vector) by means of SSR and SSS.

Apply similarity search to 269,280 s of flight data of a sortie, searching for the data of 60° bank angle turn. Search results for Threshold (DS) = 0.10 are listed in Table 4.13. WSTB indicates that in variable step-size algorithm of curve binning, the number of segments is determined in such a way that it should be an integer multiple of the sampling time. For example, when $\boldsymbol{N} = 4$, the time interval is 480 s. In Table 4.13, ratio of velocity increase refers to the ratio among WSTB and SSS and SSR, while precision refers to the ratio between the searched number and actual number.

Search results for changed of threshold values are listed in Table 4.14.

It can be seen from the search results in Tables 4.13 and 4.14 that search precision increases as the threshold value gets stricter. It is clear that threshold value deserves further research effort and may become a new research focus. The experiment results reveal that the improved multivariable similarity research method works very well in meeting various search requirements for aircraft in flight. In the application of the search method, since samples are sorted in the order of importance, and only normalized step sizes are selected for comparison, high search efficiency and precision is achieved, meeting requirements for fast search of multivariable and high-dimensional time series.

Obtaining qualified data from among a great amount of flight data lays a solid foundation to build sets of samples necessary for further analysis and research. By

Table 4.13 Search results of real flight data

Total time/s	SSS search time/s	SSR search time/s	WSTB search time/s	Velocity growth ratio
269280	5249	245	37	141.6/6.62
Number of circling	Number of SSS search	Number of SSR search	Number of WSTB search	WSTB precision
68	66	64	65	0.959

Table 4.14 Search results of real flight data

Threshold value	WSTB time/s	Number of search
0.10	37	65
0.15	25	62
0.2	14	57

taking into full consideration multivariate and high-dimensional property of flight data, changing step to set in different bins algorithm well achieves fast data search. This algorithm works on the basis of PLR, which is followed by data binning according to index tags. Inside the bins, there is a self-sequencing by the curves in terms of distance; therefore, in actual data search similar curves can be positioned by means of examining the index tag of each bin. In flight data research, variable step size method enables effective use of the original index tags. Application of this method in real data search reveals that its high speed and precision satisfies the requirements of the study of flight data.

4.6 Multivariable Flight Data Similarity Search Method Based on QR decomposition of Correlation Coefficient Matrix

In this section, QR decomposition will be used in feature extracting from correlation coefficient matrix of multivariable flight data. In addition to its low computational complexity, as well as its high efficiency and precision, this method is not restricted by time length of data series, and therefore, it can be used in similarity measuring for multivariable time series with unequal time length.

4.6.1 Representation of Matrix and Plot for Multivariable Time Series

Multivariable time series are made up of data with multiple attributes which are obtained by recording in temporal order observed values of aircraft system data of different attributes. For multivariable time series $\boldsymbol{X}$, $\boldsymbol{X}$ represents vectors in multivariable space which are derived from spanned basement of time series of different periods of time. But when number of dimensions is greater than 3, $\boldsymbol{X}$ is a high-dimensional vector and it not suitable for direct application due to its poor visualizability.

Suppose $\boldsymbol{X} = (\boldsymbol{X}_1,\ldots, \boldsymbol{X}_i,\ldots, \boldsymbol{X}_m)$ is a multivariable time series, and $\boldsymbol{Xi} = (\boldsymbol{X}_{1i},\ldots, \boldsymbol{X}_{ji},\ldots, \boldsymbol{X}_{ni})$. Here, i is the observed series corresponding to the ith period of time; n, the amount of observed series; m, the amount of attributes. If $i = 1, 2,\ldots, m$, $j = 1, 2,\ldots, n$, and $n, m \in \mathbf{N}^+$, $\boldsymbol{X}$ can be rewritten as:

$$\overline{X} = \begin{bmatrix} x_{11} & x_{12} & \cdots & x_{1m} \\ x_{21} & x_{22} & \cdots & x_{2m} \\ \vdots & \vdots & \ddots & \vdots \\ x_{n1} & x_{n2} & \cdots & x_{nm} \end{bmatrix} \tag{4.47}$$

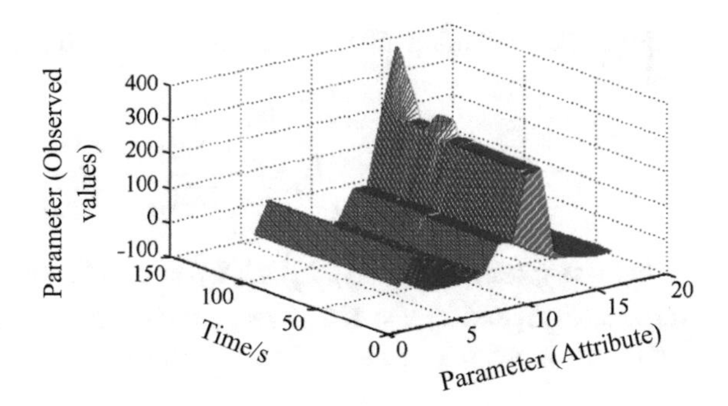

Fig. 4.37 Three dimensional representation of multivariable time series

In this expression, $\overline{X}$ is a time series matrix corresponding to multivariable time series X; each matrix row corresponds to the recorded observed values of a certain attribute, while each column represents the time series corresponding to the observed values of a certain attribute. In this time series matrix, time i, attribute j, and observed value x_{ij} work as three axes to form a three dimensional coordinate so that flight data, which is typically multivariable time series, can be represented as three dimensional graphs. Figure 4.37 is the three dimensional representation of the multivariable time series of certain aircraft.

4.6.2 *Correlative Representation of Matrix for Multivariable Time Series*

Multivariable time series reflect temporal variation of a system with respect to all its attributes, and the correlativity between attributes also reflects the correlativity inside the system. Considering the influence of inner correlativity on level of similarity of multivariable time series, there should be a correlative representation of multivariable time series on the basis of its matrix representation.

Let $\overline{X}$ be temporal matrix of time series $X = (X_1, \ldots, X_m)$, $i = 1, 2, \ldots, m$, and $m \in \mathbf{N}^+$. Due to system randomness, X_i can be seen as a random variable of time series X. As to the digital feature of the relationship between two random variables, it can be described in terms of relevant coefficients. Accordingly, the inner correlativity of multivariable time series can be represented as a correlative coefficient matrix $\widehat{X}$, referred to as correlative matrix:

$$\widehat{X} = \begin{bmatrix} \widehat{x}_{11} & \widehat{x}_{12} & \cdots & \widehat{x}_{1m} \\ \widehat{x}_{21} & \widehat{x}_{22} & \cdots & \widehat{x}_{2m} \\ \vdots & \vdots & \ddots & \vdots \\ \widehat{x}_{m1} & \widehat{x}_{m2} & \cdots & \widehat{x}_{mm} \end{bmatrix} \tag{4.48}$$

From the definition of relevant coefficients, the following expression can be derived:

$$\widehat{x}_{ij} = \frac{\mathrm{Cov}(\boldsymbol{X}_i, \boldsymbol{X}_j)}{\sqrt{D(\boldsymbol{X}_i)}\sqrt{D(\boldsymbol{X}_j)}} \tag{4.49}$$

In this expression, $\mathbf{Cov}(\boldsymbol{X}_i, \boldsymbol{X}_j)$, $\boldsymbol{D}(\boldsymbol{X}_i)$, and $\boldsymbol{D}(\boldsymbol{X}_j)$ are respectively covariance and variance of $\boldsymbol{X}_i$ and $\boldsymbol{X}_j$, $i = 1, 2,\ldots, m$, and $m \in \mathbf{N}^{+}$. It can further be seen that $\widehat{\boldsymbol{X}}$ is both a "$m \times m$" square matrix and a symmetry one, that is, $\widehat{x}_{ij} = \widehat{x}_{ji}$.

4.6.3 QR Distance Measurement for Multivariable Time Series

1. QR Decomposition of Correlative Matrix

The dimensional difference and randomness of system observation between attributes of multivariable time series give rise to strong randomness of correlative matrix. Meanwhile, the correlativity inside the system puts the attributes in pairwise correlation; therefore, without loss of generality, correlative matrix is characterized with nonsingularity. Correlative matrix is the direct reflection of correlativity inside multivariable time series.

In order to mining deeper into the correlative matrix, QR decomposition will be applied to $\widehat{\boldsymbol{X}}$ according QR decomposition principle of nonsingular matrix.

Lemma 4.1 *If A is a nonsingular matrix, there must be an orthogonal matrix* $\boldsymbol{Q}$ *and a nonsingular upper triangular matrix* $\boldsymbol{R}$. *Let* $\mathbf{A} = \mathbf{QR}$.

QR decomposition of correlative matrix can be expressed as:

$$\widehat{\boldsymbol{X}} = \mathbf{QR} \tag{4.50}$$

In this equation, orthogonal matrix $\boldsymbol{Q}$ and nonsingular upper triangular matrix $\boldsymbol{R}$ can be respectively expressed as follows:

$$\boldsymbol{Q} = (\boldsymbol{Q}_1, \ldots, \boldsymbol{Q}_i, \ldots, \boldsymbol{Q}_m) = \begin{bmatrix} q_{11} & q_{12} & \cdots & q_{1m} \\ q_{21} & q_{22} & \cdots & q_{2m} \\ \vdots & \vdots & \ddots & \vdots \\ q_{m1} & q_{m2} & \cdots & q_{mm} \end{bmatrix} \tag{4.51}$$

$$\boldsymbol{R} = \left(\boldsymbol{R}_1, \ldots, \boldsymbol{R}_j, \ldots, \boldsymbol{R}_m\right) = \begin{bmatrix} r_{11} & r_{12} & \cdots & r_{1m} \\ & r_{22} & \cdots & r_{2m} \\ & & \ddots & \vdots \\ & & & r_{mm} \end{bmatrix} \tag{4.52}$$

In these two expressions, $\boldsymbol{Q}_i$ is a column vector, and $\boldsymbol{R}_j$ a row vector; $i,j = 1,2,\ldots,m$; and $m \in \mathbf{N}^+$.

2. The Principle of Feature Extraction

According to the Givens method of QR decomposition, in the three dimensional space, orthogonal matrix $\boldsymbol{Q}$ interacts with the vectors in way of rotating geometry. Since rotation is a composition of some elementary transformations; vectors change their directions in rotation, but their modules remain unchanged. $\boldsymbol{Q}$ reveals the directionality of correlative matrix $\widehat{\boldsymbol{X}}$ in the three dimensional space, this is why orthogonal matrix $\boldsymbol{Q}$ is referred to as a directional matrix.

Make a survey of expression (4.48) through to (4.52), and the following expression can be derived from matrix multiplication:

$$\widehat{x}_{ij} = \sum_{k=1}^{m}\sum_{l=1}^{m}(q_{ki}r_{jl}) \tag{4.53}$$

In this expression, $i,j,k,l = 1,2,\ldots,m$, $m \in \mathbf{N}^+$.

An examination of (4.53) reveals that $\boldsymbol{r}_{jl}$ functions to add weighting to $\boldsymbol{q}_{ki}$. That is, by adding weighting to $\boldsymbol{Q}$, $\boldsymbol{R}$ produces scaling effect on the corresponding column vectors in $\boldsymbol{Q}$ with respect to their direction; when the scaled vectors change in their module, their direction remains unchanged. Meanwhile, $\boldsymbol{R}$ reflects the scalability of correlative matrix $\widehat{\boldsymbol{X}}$ in the three dimensional space; and therefore, it is referred to as nonsingular upper triangle weighting matrix.

Since differences between spatial vectors show themselves mainly in rotation and scalability, included angles between column vectors corresponding to $\boldsymbol{Q}$ of different multivariable time series, and the elements in $\boldsymbol{R}$ can be used as features for the measuring of similarity.

3. **QR** Distance

The discussion above indicates that the spatial differences between multivariable time series exist mainly in rotation and scalability which are influenced respectively $\boldsymbol{Q}$ and $\boldsymbol{R}$. Since similar time series should resemble to each other in rotation and scalability, the above-mentioned features can be used in measuring similarity.

For two given multivariable time series $\boldsymbol{X}^1 \in \mathbf{R}^{l\times m}$ and $\boldsymbol{X}^2 \in \mathbf{R}^{n\times m}$, there are corresponding temporal matrixes $\overline{\boldsymbol{X}^1}$ and $\overline{\boldsymbol{X}^2}$, and two correlative matrixes $\widehat{\boldsymbol{X}^1} = \boldsymbol{Q}^1\boldsymbol{R}^1$ and $\widehat{\boldsymbol{X}^2} = \boldsymbol{Q}^2\boldsymbol{R}^2$.

Directional matrixes $\boldsymbol{Q}^1$ and $\boldsymbol{Q}^2$ are expressed as follows:

$$Q^1 = (Q_1^1, \ldots, Q_i^1, \ldots, Q_m^1) = \begin{bmatrix} q_{11}^1 & q_{12}^1 & \cdots & q_{1m}^1 \\ q_{21}^1 & q_{22}^1 & \cdots & q_{2m}^1 \\ \vdots & \vdots & \ddots & \vdots \\ q_{m1}^1 & q_{m2}^1 & \cdots & q_{mm}^1 \end{bmatrix},$$

$$Q^2 = (Q_1^2, \ldots, Q_i^2, \ldots, Q_m^2) = \begin{bmatrix} q_{11}^2 & q_{12}^2 & \cdots & q_{1m}^2 \\ q_{21}^2 & q_{22}^2 & \cdots & q_{2m}^2 \\ \vdots & \vdots & \ddots & \vdots \\ q_{m1}^2 & q_{m2}^2 & \cdots & q_{mm}^2 \end{bmatrix}$$

In the matrixes, Q_i^1 and Q_i^2 are column vectors, $i = 1, 2, \ldots, m$, $m \in \mathbf{N}^+$. Weighting matrixes R^1 and R^2 are expressed as:

$$R^1 = \left(R_1^1, \ldots, R_j^1, \ldots, R_m^1\right) = \begin{bmatrix} r_{11}^1 & r_{12}^1 & \cdots & r_{1m}^1 \\ & r_{22}^1 & \cdots & r_{2m}^1 \\ & & \ddots & \vdots \\ & & & r_{mm}^1 \end{bmatrix},$$

$$R^2 = \left(R_1^2, \ldots, R_j^2, \ldots, R_m^2\right) = \begin{bmatrix} r_{11}^2 & r_{12}^2 & \cdots & r_{1m}^2 \\ & r_{22}^2 & \cdots & r_{2m}^2 \\ & & \ddots & \vdots \\ & & & r_{mm}^2 \end{bmatrix}$$

In the matrixes, R_j^1 and R_j^2 are row vectors, $j = 1, 2, \ldots, m$, and $m \in \mathbf{N}^+$.

A survey of expression (4.48) and (4.50) shows that correlative matrix $\widehat{X}$ is a square matrix, and its upper and lower triangle elements are symmetrical and equal in terms of diagonal line, namely $\widehat{x}_{ij} = \widehat{x}_{ji}$, thus, $\widehat{x}_{ij}$ and $\widehat{x}_{ji}$ are determined by Q_i and R_i.

Definition 4.11 QR distance can be used as similarity measurement between multivariable time series $X^1 \in R^{l \times m}$ and $X^2 \in R^{n \times m}$:

$$D_{\text{QR}}(X^1, X^2) = \sum_{i=1}^{m} \lambda_i [1 - \cos(\varphi_i)] \tag{4.54}$$

In this expression, $\lambda_i = \sum_{j=1}^{m} \left(r_{ji}^1 + r_{ji}^2\right) \Big/ \sum_{i=1}^{m} \sum_{j=1}^{m} \left(r_{ji}^1 + r_{ji}^2\right)$, therefore, $D_{\text{QR}}\left(X^1, X^2\right) \in [0, 1]$, $i, j = 1, 2, \ldots, m$, and $m \in \mathbf{N}^+$. $\varphi_i = \left\langle Q_i^1, Q_i^2 \right\rangle$ is the vectorial angle between spatial vectors $\boldsymbol{Q}_i^1$ and $\boldsymbol{Q}_i^2$. From the cosine formula of vectorial angle the following can be derived:

$$\cos(\varphi_i) = \cos\left(\langle \boldsymbol{Q}_i^1, \boldsymbol{Q}_i^2 \rangle\right) = \frac{(\boldsymbol{Q}_i^1, \boldsymbol{Q}_i^2)}{|\boldsymbol{Q}_i^1||\boldsymbol{Q}_i^2|} \tag{4.55}$$

In this expression, $(\boldsymbol{Q}_i^1, \boldsymbol{Q}_i^2)$, $|\boldsymbol{Q}_i^1|$ and $|\boldsymbol{Q}_i^2|$ represent respectively the inner product and module of $\boldsymbol{Q}_i^1$ and $\boldsymbol{Q}_i^2$.

A survey of expression (4.54) and (4.55) reveals that if a directional matrix is at same time s square one, both spatial vectors $\boldsymbol{Q}_i^1$ and $\boldsymbol{Q}_i^2$ are unit vectors, namely, $|\boldsymbol{Q}_i^1| = |\boldsymbol{Q}_i^2| = 1$. Therefore, **QR** distance can be expressed as follows:

$$D_{\mathrm{QR}}\left(\boldsymbol{X}^1, \boldsymbol{X}^2\right) = \sum_{i=1}^{m} \lambda_i \left[1 - (\boldsymbol{Q}_i^1, \boldsymbol{Q}_i^2)\right] \tag{4.56}$$

4.6.4 Verification on Clustering of Multivariable Flight Data Based on QR Distance

4 groups of 100-frame (frame/s) continuous flight data of certain aircraft are chosen for experiments. Each group consists of 15 types of flight data, making 4 multivariable time series on a 100 × 15 scale, namely, $\boldsymbol{X}^1$–$\boldsymbol{X}^4$ as shown in Fig. 4.38 (The vertical axis data have been technologically processed, and will not affect analysis of experiment results.). The flight data cover acceleration of axial direction, lateral direction, normal acceleration, rotor speed of left and right engine, exhaust temperature, as well as torque.

Experiment 1: Verification of effectiveness and precision of QR distance. As shown in Table 4.15, QR distance between $\boldsymbol{X}^1$ and $\boldsymbol{X}^4$ in pairs, Euclidean distance, and SVD distance are calculated. For sake of an easy comparison, and for avoidance of effect on experiment results, unified magnitude processing has been done to the calculation results of SVD distance. Following a comprehensive survey of the data size and graphic patterns of each time series in Fig. 4.38, and an analysis in reference to Table 4.15, it can be seen that both QR and SVD distance, which take into account the inner correlativity of multivariable time series, conclude the highest level of similarity between $\boldsymbol{X}_1$ and $\boldsymbol{X}_4$. This conclusion agrees with manual interpretation. Euclidean distance does not take into account the inner correlativity, and concludes the highest level of similarity between $\boldsymbol{X}_1$ and $\boldsymbol{X}_2$. This conclusion contradicts manual interpretation, and has obvious limitations. All the three types of distance agree that $\boldsymbol{X}_2$ and $\boldsymbol{X}_3$ share the least similarity among them, which agrees with manual interpretation. This experiment verifies that QR distance is a highly effective and precise measurement for similarity between multivariable time series.

Fig. 4.38 Three dimensional graphs of time series of 4 groups multivariable flight data of equal time duration

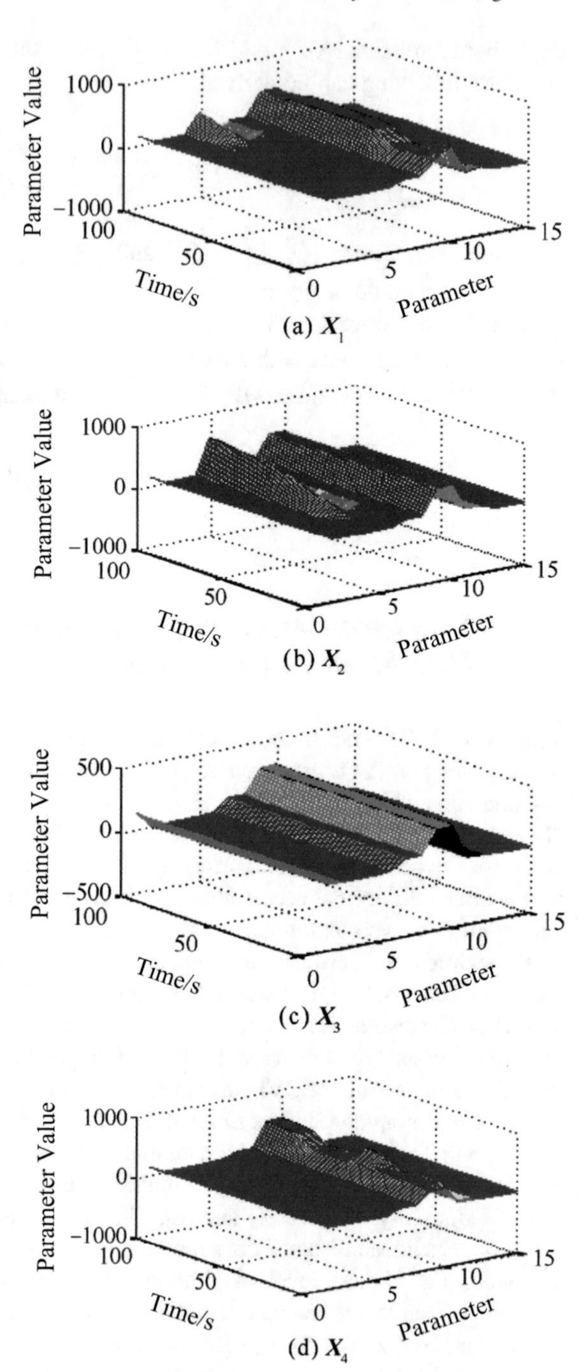

Table 4.15 Calculation results of three types of distance

Type	$D_{QR}(X_1, X_2)$	$D_{QR}(X_1, X_3)$	$D_{QR}(X_1, X_4)$	$D_{QR}(X_2, X_3)$	$D_{QR}(X_2, X_4)$	$D_{QR}(X_3\ X_4)$
QR	0.4621	0.6036	0.3787	0.6463	0.4414	0.4792
Euclidean	1.2238	2.6100	1.7284	2.6207	1.7189	2.2364
SVD	2.1467	8.3267	1.5529	9.5190	7.2511	2.2898

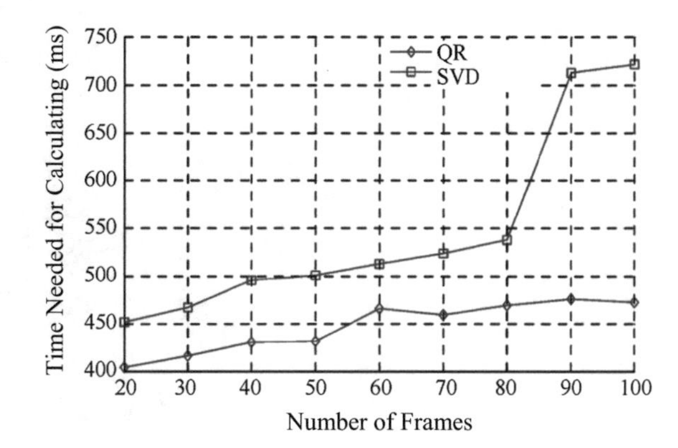

Fig. 4.39 Comparison of calculation time

Experiment 2: Verification of higher effectiveness of QR distance to SVD distance. Respectively in multivariable time series $\boldsymbol{X}_1$ and $\boldsymbol{X}_2$ in Fig. 4.38, from the 10th frame, and at an interval of 10 frames, superimpose one by one to make 9 new time series of unequal time duration. Then, calculate the time needed by the two distance algorithms in measuring the distance between the paired new series on the same scale. The calculation results are shown in Fig. 4.39.

It can be seen from Fig. 4.39 that in all the calculations of distance of similarity in the same time series, and on the condition that effectiveness and precision are ensured, QR distance costs less time than SVD algorithm; and that when time series grow in scale, time needed for SVD fluctuates in a big way while QR shows very small change in time consumption.

Experiment 3: Verification of QR algorithm with respect to its advantage of not being restricting by time duration of time series. A survey of expression (4.51) and (4.52) shows that both directional and weighting matrix are square matrixes the scale of which is determined by number of columns, namely the number of attributes possessed by a multivariable time series, and has nothing to do with time duration of the series. According to the definition of QR distance, it can be used to measure the similarity of the time series with the same amount of attributes but different time length. Choose for verification 4 groups of multivariable time series time durations of which are respectively 100, 200, 300, and 400 frames (frame/s). As shown by Fig. 4.40 (The vertical axis data have been technologically processed, and will not affect analysis of experiment results.), the 4 times series are expressed respectively as $\boldsymbol{X}_1$–$\boldsymbol{X}_4$.

Fig. 4.40 Three dimensional graphs of time series of 4 groups multivariable flight data of unequal time duration

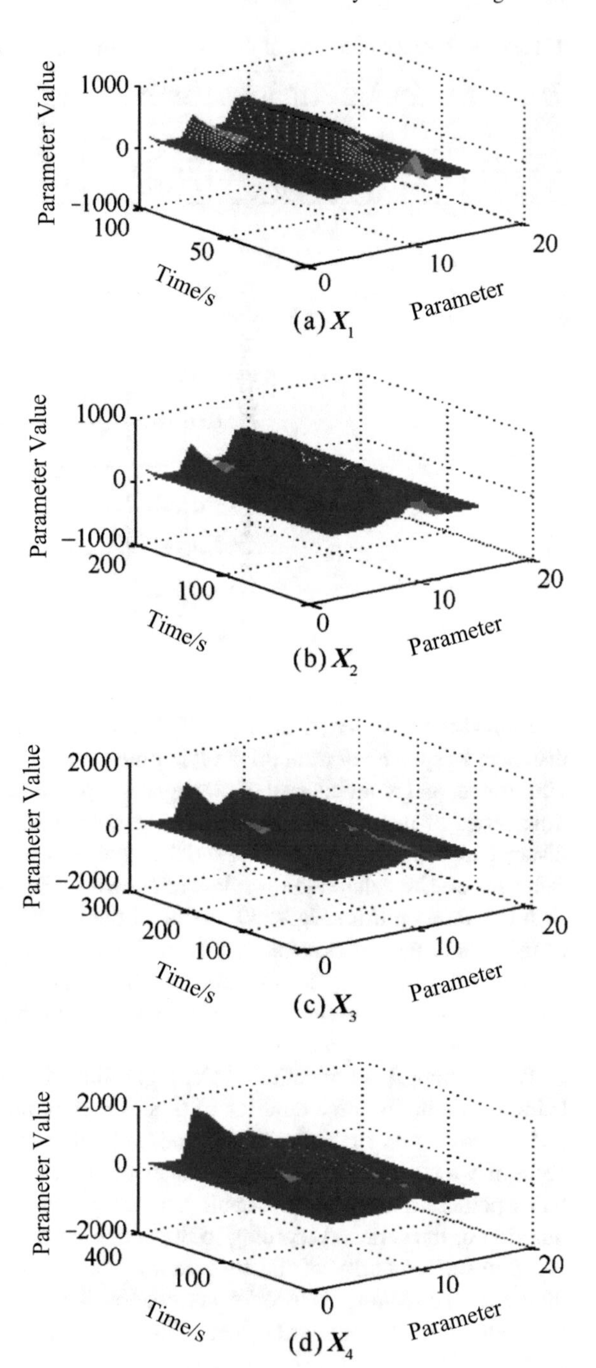

Table 4.16 Calculation results of QR distance between multivariable time series of different time duration

Type	$D_{QR}(X_1, X_2)$	$D_{QR}(X_1, X_3)$	$D_{QR}(X_1, X_4)$	$D_{QR}(X_2, X_3)$	$D_{QR}(X_2, X_4)$	$D_{QR}(X_3, X_4)$
QR	0.3298	0.3779	0.3557	0.1287	0.0972	0.0185

Calculate QR distance between multivariable time series of unequal time durations, namely X^1–X^4 in pairs; the calculation results are listed in Table 4.16. After a comprehensive survey of the data size and graphic patterns of each time series in Fig. 4.38, and an analysis in reference to Table 4.16, results of QR distance concludes the highest level of similarity between X_3 and X_4, but the lowest level of similarity between X_1 and X_3, which agrees with manual interpretation, and has obvious limitations. This experiment verifies that QR distance is a highly effective and precise measurement for similarity between multivariable time series with unequal time durations.

Chapter 5
Condition Monitoring and Trend Prediction Based on Flight Data

One important purpose of intelligent processing and application analysis of flight data is to conduct daily monitoring and evaluation of flight quality and aircraft performance so as to provide a sound basis for ground maintenance and flight training. This chapter is an introduction to the methods of aircraft condition monitoring, and an elaboration of the diagnostic methods of gradual and abrupt faults based on expert system and dynamic principle component analysis. The monitoring methods introduced here are based on flexible-size grid technology, weighted least squares support vector machine (WLS-SVM), and chaos theory. Both the monitoring and diagnostic methods are verified by using real data.

5.1 Clustering Methods Based on Flexible-Size Grid for Airplane Equipment

The processing of flight data often finds itself in a situation of learning without teachers, that is, unsupervised learning. If the expert neither specifies the condition represented by the data, nor provides the specific rules and threshold values, it is necessary for data processing personnel to have methods that are methods capable of automatic clustering analysis of the data. Currently in the research of clustering analysis, self-organizing neural network method enables automatic clustering analysis by simulating the working of human brain, but classification of the final results depends highly on manual labor.

The purpose of condition monitoring of aircraft equipments is to get a clear picture of their performance. The airframe movement equation obtained through modeling can only roughly reflect the changes of the airframe as a rigid body, and cannot evaluate aircraft performance. As a result, the central issue for the present study is to find from the well-developed methods a suitable one for clustering analysis of flight data. Generally, clustering analysis makes use of a great number of

J. Zhang and P. Zhang, *Time Series Analysis Methods and Applications for Flight Data*, DOI 10.1007/978-3-662-53430-4_5

unmarked samples in its automatic training of the classifier, but manual marking of the data groups may yield different results. In this section, it is proposed that in flight data processing, different two-dimensional data densities can be acquired by adopting the method of flexible-size grid, and that automatic data scaling can be realized by means of monolithic translation. Effective clusterings have been obtained in this way, and analysis of the engine data shows that the method proposed works well in automatic clustering analysis.

5.1.1 Clustering Methods of Multivariate Data

With the increase of data dimensions, generally very few dimensions of the data are related some of the clusters, but the data of those unrelated dimensions may produce much noise which shields the real clusters. Furthermore, since the data will generally become sparser, the majority of the data points will be distributed in the subspace of different dimensions. Where there are very sparse data, data points of different dimensions may be considered to be in equal distance from each other, and consequently, distance measurement, which is a very important element in clustering analysis, will lose its importance.

1. Clustering method of dimension increasing subspace

CLIQUE is the first space clustering algorithm of dimension increasing subspace in high-dimensional space. In the clustering of dimension increasing subspace, the clustering process starts from a single-dimension subspace and moves on into subspaces of higher dimensions. Since CLIQUE divides each dimension into grid structure, and determines whether the grid cell is dense or not according to the number of the points included in each grid cell, it can also be perceived as an integration of two types of clustering, that is, density-based and grid-based.

The principle of CLIQUE clustering algorithm can be summarized as follows. First, a big set of the multidimensional data points is established, and the data points are often not evenly distributed in the data space. In locating the sparse and the crowded areas in the data space by means of CLIQUE, the overall distribution pattern of the data set can be obtained. Second, if a cell is dense, it means that the total number of the included data points exceeds that of input parameter of a certain model. In CLIQUE, a cluster is defined as the maximum set of the interconnected dense cells.

CLIQUE is not sensitive to the order of the input objects. When the input size scales linearly and the number of dimension increases, there will be good scalability. However, the accuracy of the clustering results may decline at the expense of algorithmic complexity. In addition, for a given dense area, its projection of this area in all low-dimensional subspaces will be dense, which may result in much overlap in the reported dense areas, and consequently, it will be difficult to identify in subspaces of different dimensions the clusters which are different in density.

2. Clustering method for subspace dimension reduction

PROCLUES is a typical clustering method for subspace dimension reduction. Clustering does not start from a single dimension, but from searching for the cluster's initial approximation in the high-dimensional attribute space. Every cluster in each dimension is assigned a weight which is used in the next round of iteration to create a new cluster. All subspaces of the expected dimension are detected for dense areas, and in doing so, excessive overlaid clusters should be avoided in the projection areas in lower-dimension subspaces.

The algorithm of PROCLUS is divided into three phases, namely, initialization, iteration, and clustering improvement. In the initialization phase, the greedy algorithm is used to select a group of initial center points which are far away from each other and it should be made sure that each cluster should be represented at least by one of these points. To be specific, first of all, it proportionately selects the samples of the data points at random according to the number of clusters to be created. Then, a smaller subset, which is to be used in the next phase, is acquired by using the greedy algorithm. In the iteration phase, k center points are selected randomly in set after reduction. If the clustering quality is to be improved, new points would be randomly selected to replace those center points which are "not good". For each center point, the set of the dimensions whose average distance is shorter than the expected value should be selected.

3. Clustering method based on frequent pattern

The idea that underlies clustering analysis based on frequent pattern is that frequent pattern may predict the emergence of a cluster. This method can be perceived as an expansion of the clustering method of dimension increasing subspaces. The boundary between different dimensions, however, is not very obvious, because they are expressed in terms of the set of frequent item sets. That is to say, the description of clustering is not obtained through dimension-by-dimension clustering growth, but by the growth of sets of frequent items. The typical example of the analysis based on frequent pattern includes the clustering of text documents containing thousands upon thousands of key words. Figure 5.1 is an example of the analysis of microarray data containing thousands upon thousands of measurement values or "characteristics".

5.1.2 Clustering Method Based on Density Function

In order to solve the problem of how to find the cluster of random shape, the clustering method based on density function is developed. Generally, in this method, cluster is seen as the dense object areas which are separated by low-density areas in the data space. The basic thinking behind density clustering generally concerns neighborhood, key objects, direct density-reachablility, and density connectivity. In addition, the density-based cluster is the maximum set of

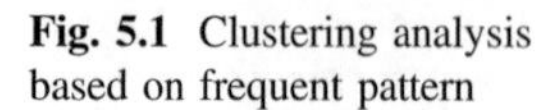

Fig. 5.1 Clustering analysis based on frequent pattern

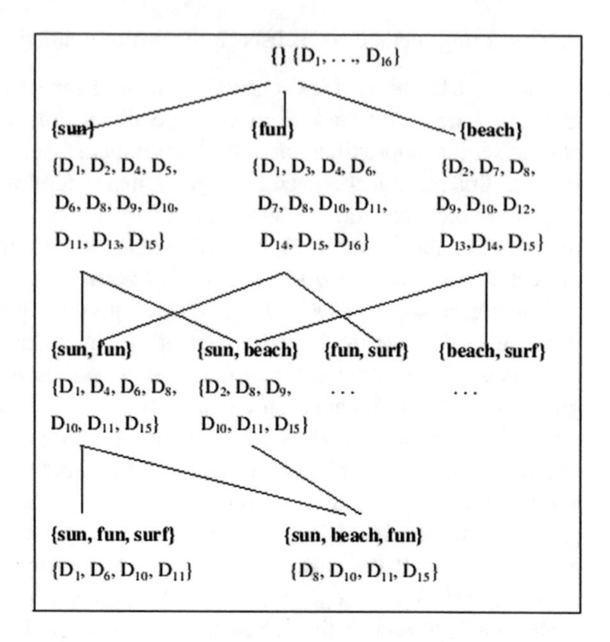

density-connected objects based on density reachability, and does not include the objects which are considered as noise in any cluster.

1. Density-based spatial clustering of applications with noise: DBSCAN

DBSCAN is a clustering algorithm based on density. According to this algorithm, the space whose density is high enough is divided into clusters. Clusters of random shapes in the space database with noise will be identified, and cluster is defined as the maximum set of the density-connected points. DBSCAN searches for clusters through examining the neighborhood of every point in the database. If the number of points included in the neighborhood of this point is greater than a specified number, then a new cluster of the object is created, and the objects which can be directly density reachable from these key objects are iteratively aggregated. This process may involve the merging of some density reachable clusters. The whole process will be terminated when no new point can be added to any cluster. Figure 5.2 illustrates the definition of density reachability and connectivity.

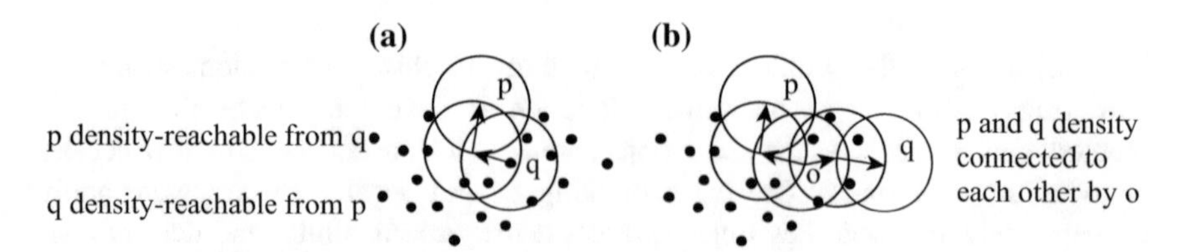

Fig. 5.2 Density reachability and connectivity in density-based clustering

2. Ordering points to identify the clustering structure: OPTICS

OPTICS does not directly produce data set clustering, but calculates one augmented cluster ranking for automatic and interactive clustering, which represents the clustering structure of the data based on density. OPTICS algorithm creates the ranking of subjects in the database, and additionally saves the core distance and relative reachable distance of every subject. For those whose distance is shorter than this ranking, all the clustering based on density is extracted. Figure 5.3 is a demonstration of the definition of core distance.

3. Density-based clustering: DENCLUE

DENCLUE is a kind of clustering algorithm based on a group of density distribution function. The idea of this algorithm is that the influence of every datum can be formally modeled by using a mathematical function. This function is called influence function, which mainly describes the influence of data points in their neighborhood. The overall density of the data space can be modeled by using the sum of the influence function of all the data points. Clusters can be determined by identifying the number of the density attractive points, which are the local maxima of the overall density function. Figure 5.4 shows the density of 2D data sets.

5.1.3 Shrinking Clustering Method Based on Flexible-Size Grid

This section provides the flexible-size grid shrinking clustering method (FG-SCM). The advantages of this method are as follows. First, it can automatically select the size of the grid. Second, the result of calculation is acquired when the shrinking has finished. Finally, the results of clustering to be received are dependent upon the actual structure of the data, which ensures the validity and accuracy of the result.

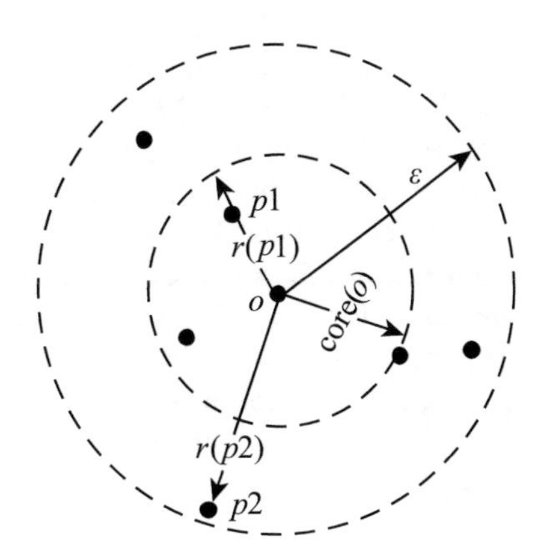

Fig. 5.3 The core distance in OPTICS

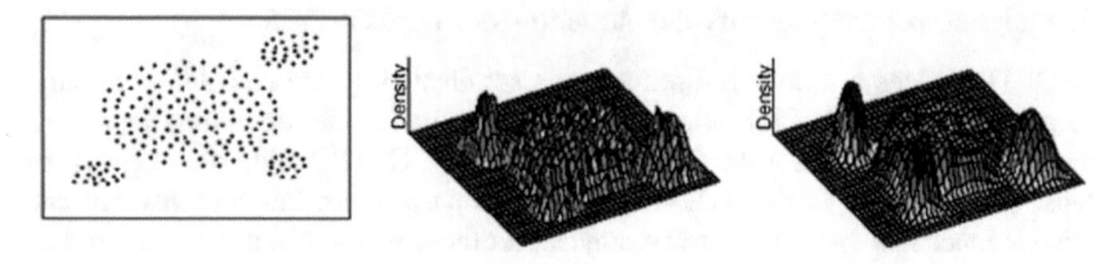

Fig. 5.4 The possible density function of 2D data sets

1. The principle of choosing grid size

Grid clustering method sets high requirement for the size of the grid cell. It is very difficult to choose the correct grid size if the structure of the input data is not known in advance. Following data marshaling by means of flexible-size grid, shrinking and clustering test is done to the data, and the grid size can be determined by choosing the optimal test result.

Suppose the input data set is $\boldsymbol{X} = \left\{\overrightarrow{X_1}, \overrightarrow{X_2}, \ldots, \overrightarrow{X_n}\right\}$, it can be regularized into hypercube as $[0, 1)^d \subset \mathrm{IR}^d$, then the reasonable size of shrinking clustering can be acquired by using column map. After one-time search of the input d dimensional data, the column map can be acquired, with each column map representing the data of one dimension. In $\boldsymbol{H} = \{h_1, h_2, \ldots, h_d\}$, each item represents the number of the item in a specific distance in the column map. Set β as the threshold of the number, and calculate the density gradient.

Definition 5.1 Density span is the combination of continuous data box when the number of the points in a certain distance is larger than β. It is the number of the boxes included.

For each histogram, the data boxes received are sorted in descending order according to the number of the data. Then, starting from the first data box, the data boxes are combined one by one until the overall number is larger than β. The data on both sides of the box are examined and combined to get the relevant density span. If the overall data number in this span is larger than β, but the two adjacent points have already been set to one previous span, this process would be terminated, and it is called unfinished span, and will not be considered in the following selection of flexible-sized grids.

As it is shown in Fig. 5.5, starting from box 21, the adjacent two boxes—box 20 and box 22—are checked. Box 20 is selected to be combined because the number of the points in this box is larger than that in box 22. The box 19 and box 22, which are beside box 20 and box 21, are selected. Box 19 is select to be combined. The neighboring boxes are combined and finally the set number is reached and surpassed, and the resulting data span is 1. Data box 7 is the box whose number of the data points is the second largest. Density span 2 can be got, which includes 7, 8, 6, 9, 10 and 5.

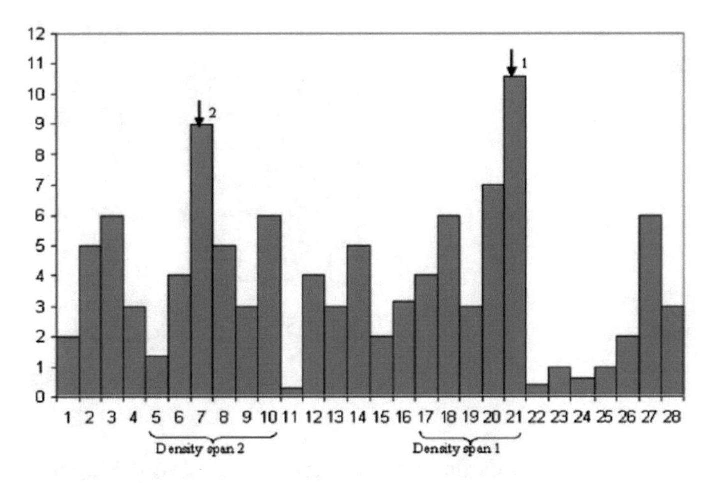

Fig. 5.5 Method to acquire density span

The density span is clustered according to the size. The specific procedure is provided as follows. First, sort the size in the ascending order, and start to set cluster T_1 for the smallest size S_0. If in size S', $(S' - S_0) > S_0 \times 5\%$, start to set T_2 for S'. Then, mark the number N_i for each cluster, and mark the mean value S_i. Rank the data in the descending order. K_s is selected as the multidimensional scaling to be used in data shrinking clustering and clustering differentiation process, and the density span is finally acquired. The following algorithm is the specific procedures for producing density span.

The algorithm for calculating density span:

Input: histogram parameter h_i
Output: the density span set for h_i
Step 1: sort h_i in the descending order;
Step 2: starting from the first box of the ranked box set, combine it with its neighboring two boxes until the number of the data points is larger than threshold β;
Step 3: repeat step 2 and stop when all the boxes which are not empty are combined;
Step 4: output density span set.

The setup of the value for β is related to the input data set X. The value of β can generally be set as the specific ratio in $\boldsymbol{X}$. A smaller K_s would increase the accuracy of clustering, while a bigger one would save time. Time complexity can be determined by the amplitude d of X and the number of the boxes B_n in each histogram. The execution time for this algorithm is $O(B_n)$ lg B_n.

The method for selecting grid in multidimensional grids can not only determine the size of the cell, but also provide the validity of data sets with different densities. In Fig. 5.6, there are three types of data. The densities of the two types of data on the left are higher than that on the right. The smaller cell grids can be used to

Fig. 5.6 Data sets divided into three types

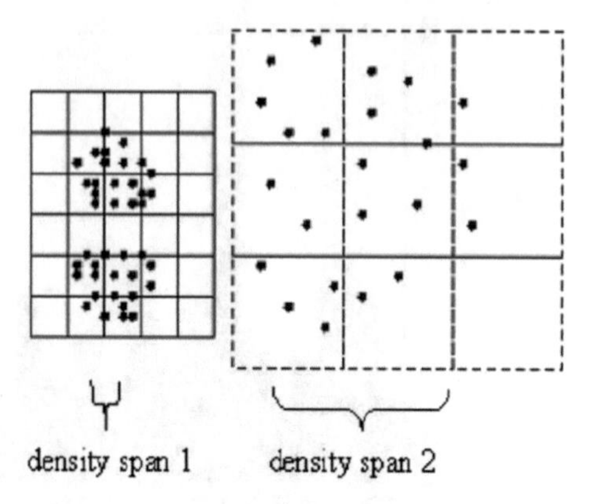

differentiate the two types on the left, but cannot be used to differentiate the right one. The contrary is the case for the big grids, so flexible-sized grids are needed to differentiate all the patterns.

2. The method for flexible-sized grids

Moving along the gradient direction, the data points are gradually shrinking inwards, and then they are pulled by neighboring data to create denser clusters. The relationship between neighboring points and data sets is formed by grids. The space is first divided into grid cells. The points in sparse cells are perceived as noise fields or outliers, and are neglected in the shrinking process. Suppose density cell C is surrounded by neighboring cells with higher density, the data will move towards the neighboring cells until the preset conditions are met. Neglect of the moving of sparse points can save the calculating time: if the girds are small enough, the nonzero cell may be $O(n)$, and the time for calculating the moving data is $O(n^2)$. Hypercube $[0,1)^d$ can be divided into K^d cells:

(1) The method for subdividing space

The accuracy of the histogram can be improved by using flexible-sized grids. Suppose the girds are divided into k cells, then it can be written as:

$$\left\{C(i_1, i_2, \ldots, i_d) = \left[\frac{i_1}{k}, \frac{i_1+1}{k}\right) \times \left[\frac{i_2}{k}, \frac{i_2+1}{k}\right) \times \cdots \times \left[\frac{i_d}{k}, \frac{i_d+1}{k}\right) \right.$$
$$\left. |i_1, i_2, \ldots, i_d \in \{0, 1, \ldots, k-1\}\right\}. \tag{5.1}$$

Every cell in it has a peculiar ID: $(i_1, i_2, \ldots, i_d)$. The condition for neighboring cells is $|i_k - j_k| \leq 1,\ k = 1, 2, \ldots, d$. The cell including every data points can be marked as Cell(X_i). When the data is input into the cell, the nonzero cells and other points included in them can be found. Then, mark the density of each nonzero cell on each cell. Each cell is perceived as sparse or dense based on whether its density has surpassed T_{dn1}.

Definition 5.2 Density cell can be written as

$$\textbf{\textit{DenseCellSet}} = \{C_1, C_2, \ldots, C_m\} \tag{5.2}$$

Definition 5.3 The centroid for all the points of each cell can be written as

$$\textbf{\textit{DataCentroid}}\ (C) = \frac{\Sigma_{j=1}^{k} \overrightarrow{X_{i_j}}}{k} \tag{5.3}$$

$\left\{\overrightarrow{X_{i_j}}\right\}_{j=1}^{k}$ is the marking method for all the points in this cell, and is called the data centroid of C. Each cell has its own data points and data centroid. The calculating process includes density cell, data points and centroid. The calculating time is $O(n \lg n)$, and the number of cells is $O(n)$.

In high-dimensional space, problems may arise if sparse cells are neglected. Figure 5.7 shows the different clustering methods of four points in two-dimensional space. In Fig. 5.7a, the four data points are around one grid point. It is possible that 2^d similar data may exist in d dimensional Euclid space IR^d, and the shrinking of the data may be influenced. While in real process, they may be neglected because they exist in different cells (for example Fig. 5.7b). In order to solve this problem, cross grid (Fig. 5.7c) is used for one given cell. If flexible-size grid is used, it is possible that under any circumstance, dense grids may be acquired, and it cannot be used in real calculation. Now two types of grids with different scales are used, and the second grid is produced by using the diagonal line of the first grid. Similarly, diagonal line flexible-size grid is a supplement to the first grid.

The advantages of this method are as follows: first, the structure of the cluster can be researched from different grids; second, it can be related to the method of density span which has been mentioned Fig. 5.7b; third, compared with the methods of free selection of scale (with 2^d options), the possibility of scale change is comparatively smaller.

(2) The terminating method for data shrinking

Ideally, if the data set has many types of boundary shapes, the main shapes can be obtained following the shrinking of the boundary points. If these shapes still

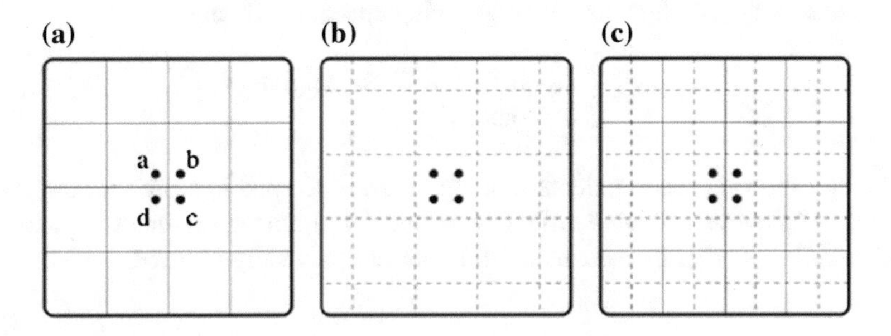

Fig. 5.7 Different clustering methods of four points in 2D space

have many types of boundary, they will continue shrink, until no new type of boundary line is obtained.

Suppose in step i, the movements of n data points are, respectively, $v_1, v_2, \ldots, v_n$, then the mean movement will be:

$$\frac{\Sigma_{j=1}^{n} \left\| \vec{v}_j \right\|}{n} \tag{5.4}$$

If the continuous movement is smaller than the threshold T_{amv}, the data set would be seen as stable shrinking, and the algorithm will be terminated. In order to effectively control the overall calculating time, an upper limit T_{it} is set for the iteration time.

(3) The analysis of time and space of calculation complexity

In the process of data shrinking, the location of every point needs to be retained, and the summation is $O(n)$. The data points are allocated into each grid, and each data cell includes all the data points, centroids, and movements. The space taken by data structure is O_n, and the time for each step is $O(m^2 + n \lg n)$, in which m is the number of the cell. The overall time is $O[T_{\mathrm{it}}(m^2 + n \lg n)]$ when the iteration procedure is set as T_{it}, and M here refers to the cell with the highest density in all procedures.

In order to control M, the lower limit of density is set as T_{dn1}, and the number of the data points in density cell should not be more than:

$$\frac{n}{T_{\mathrm{dn1}} \times \left(\frac{1}{k}\right)^d} \leq \tilde{M} \tag{5.5}$$

When the density is lower than T_{dn1}, this method is not valid, and all nonzero cells may be ranked according to the data points. If only 1 or 2 points are in each cell, bigger scale is needed.

The result of the shrinking data is examined by using the examining method based on grid clustering. The clustering examining method is conducted in crossly used grids. Suppose DC_1 and DC_2 are two groups of overlapping data grids, when $C_1 \cap C_2 \neq \varphi$, the two cells, $C_1 \in \mathrm{DC}_1$ and $C_2 \in \mathrm{DC}_2$, are called neighboring cells. Suppose E is the combination of neighboring space, in which

$$E_{ij} = \begin{cases} 1, & \text{if } C_i \text{ and } C_j \text{ are neighbors} \\ 0 & \text{otherwise} \end{cases} \tag{5.6}$$

In the following section, depth-first algorithm is adopted to search for the graph elements. As a result of the use of the shrinking algorithm whose time complexity is $O(\mathrm{DC}_1 \cup \mathrm{DC}_2 + |E|)$, the calculation time needed is greatly reduced.

3. The calculation and selection of clustering

According to the definition of good clustering, the algorithm to judge clustering validity is conducted through the judgment of the quality of all the clustering. The intelligence clustering judgment technique can be used to examine single clustering. The relationship between every point of the clustering and every point in the type is comparatively close. The relationship within the type is measured by its tightness, while the relationship outside the type is done by its separability. This definition is used to calculate the examination of clustering with different scales.

The search for shrinking subset in the graph can be simplified as the search for the minimum spanning tree (MST). Suppose $G = \langle V, E \rangle$ is a fixed pattern, V is the setup for top point, and E is the setup for boundary, then the Internal Connecting Distance (ICD) and External Connecting Distance (ECD) are adopted to measure the internal tightness and external tightness. Tightness is the ratio between external connecting distance and ICD.

Definition 5.4 For graph $G = \langle V, E \rangle$, positive weight function w is given, and there is connected subset $T \subseteq V$. Suppose $L = \{w(p, q) | (p, q) \in E, p \in T, q \notin T\}$, then the ECD in T is defined as: ECD(T; G, w) = min L. If L is empty, then ECD(T; G, w) is defined as infinity.

In the above definition, min L marks the minimum number of set L.

Definition 5.5 For graph $G = \langle V, E \rangle$, positive weight function w is given, and there is connected subset $T \subseteq V$. Suppose $L = \{l \in \mathrm{IR}^+ < T, |(p, q) \in E | p \in T, q \notin T, w(p, q) \leq 1\}$ is the connected subgraph of G, define the ICD of T as $\mathrm{ICD}(T;\ G,\ w) = \inf L$, in which inf L marks the maximum boundary for L.

$\mathrm{ECD}(T;\ G, w)$ is the shortest boundary connecting distance of T and $V - T$. $\mathrm{ICD}(T;\ G, w)$ is the shortest distance to ensure internal connection. It can be seen that $\mathrm{ECD}(T;\ G,\ w)$ is infinite when T is the connecting element of G, but for any connecting subset T, $\mathrm{ICD}(T;\ G, w)$ is finite. The connected subgraph $\langle T, H \rangle$, thus the weight of all the boundary line in H will not be more than $\mathrm{ICD}(T;\ G, w)$.

Definition 5.6 For graph $G = \langle V, E \rangle$, positive weight function w is given, and there is connected subset $T \subseteq V$. The compactness of T can be defined as:

$$\text{Compactness }(T;\ G, w) = \frac{\mathrm{ECD}(T;\ G, w)}{\mathrm{ICD}(T;\ G, w)} \tag{5.7}$$

When the ratio of compactness is over 1, it is compacted. Define all the set in G as Cluster Set (G, w).

The following algorithm Cluster Finding (G, w) is the searching method for connecting elements.

Input: $G = \langle V, E \rangle, w : V \to R^+$ //the graph curve parameter
Output: C //all the clustering result

Step 1. Set $D = \{\{p\}|p \in V\}$, and let there be function $ecl : D \to R^+$, $icl : D \to R^+$

Step 2. rank E in ascending order according to the weight of all the boundaries;

Step 3. if $p \in P \in D, q \in Q \in D$ and $P \neq Q$, divide the specific points of the clustering location for each edge$(p, q) \in E$;

Step 4. set $ecl(P) = ecl(Q) = w(p, q)$,
if $ecl(P) > icl(P)$, $C = +P$
and if $ecl(Q) > icl(Q)$, $C = +Q$
then, remove P, Q, $D = +P \cup Q$; //after getting the result, confirm and remove the scattered points;

Step 5. set $icl(P \cup Q) = w(p, q)$, for each $X \in D$ set $ecl(X) = \infty$, $C = +X$ //confirmed connected elements;

Step 6. Output C.

Definition 5.7 For the connecting subset S of V, first set MST(S) as the minimum produced tree including S minimum subgraph. The ICD of S is marked as ICD(S; G, w), which is the longest boundary distance of MST(S). While the ECD is marked as ECD(S; G, w), which is defined as including the shortest boundary connection S and V–S. S is called as compacted point set, whose compactness is defined as:

$$\text{Compactness}\ (S;\ G, w) = \frac{\text{ECD}(S, G, w)}{\text{ICD}(S, G, w)} \tag{5.8}$$

It is the calculation of dataset in low-dimensional Euclidean space. Here Delaunay graph is adopted.

4. The application area of algorithm

Delaunay graph cannot be effectively constructed in high-dimensional space. The compactness of the two clusters in Fig. 5.8 may be affected by noise. The actual compaction of this algorithm must be conducted on the basis of the whole graph. Suppose only the data in the left and middle parts of Fig. 5.8 are compacted, then only can the racket-shaped result be obtained, and the noise data would be retained. But if the data in the whole graph is compacted, two clustering results can be obtained and the noise in the middle can be eliminated.

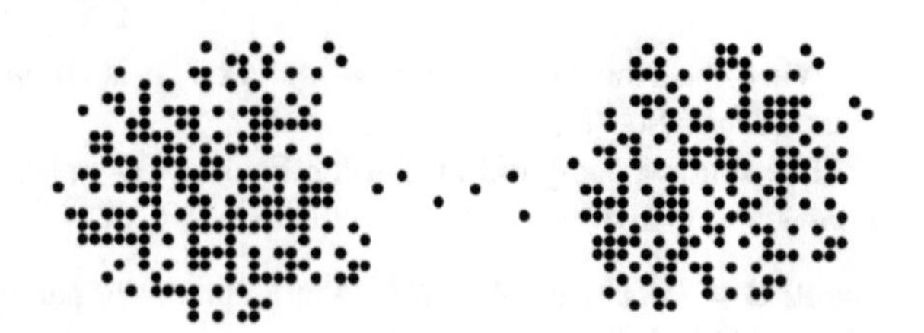

Fig. 5.8 Two sets of data with noise points

5.1.4 Monitoring Examples of Flight Equipment Status Shrinking Clustering

Self-organizing neural network can quickly map the multidimensional data onto two-dimensional plane. If there are lots of data points, manual clustering work is still needed for the reduced-dimensional data, which makes this method bearing considerable limitation. If manual drawing method is adopted to cluster the data, there may be mistakes in the result. Therefore, solving the self-clustering problem in two-dimensional plane will make this method more practical. Figure 5.9 is the flowchart of engine SOFM shrinking clustering method.

Monitoring the status by adopting engine flight data have the following advantages: (1) engine has comparatively typical faults, and the parameters in flight data recording system can comparatively describe its working condition completely; (2) the corresponding parameter in flight data can be obtained during the intervals of the flight mission; (3) the historical database of the flight data can serve as a stable data source for engine performance monitoring.

1. The standardization of flight data

The flight data need to be standardized before they are input into SOFM network. The standardization of input value of the attribute measurement of the training samples can help to quicken the neural network learning phase. The attribute data is shrunk pro rata, and the raw data is changed linearly by using minimum–maximum standardization. Among them, v' is the value changed, v is the actual value of the data, max A is the maximum value of the original measurement, new_max_A is the maximum value of the new measurement, and new_min_A is the minimum value of the new measurement:

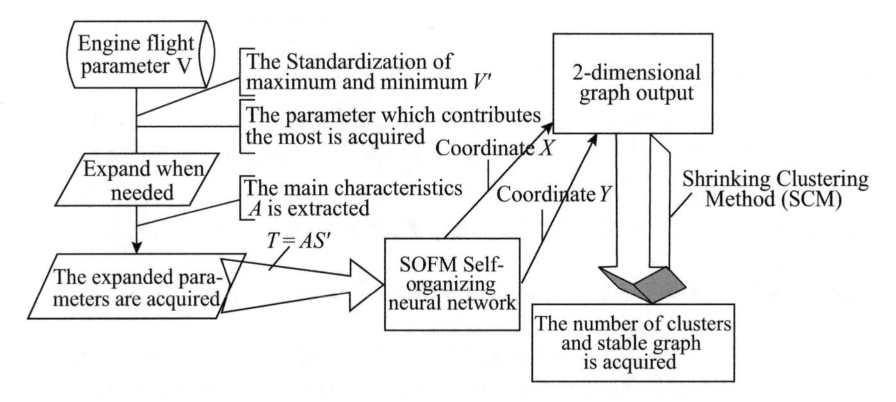

Fig. 5.9 The flowchart of engine SOFM shrinking clustering method

$$v' = \frac{v - \min A}{\max A - \min A}(new\_\max_A - new\_\min_A) + new\_\min_A \tag{5.9}$$

Change the value of A into interval [0, 1], that is, $v' \in [0, 1]$, and then map it onto the v' in $new\_\max_A, new\_\min_A$. The standardization of maximum–minimum maintains the relationship of the raw data. The input data should not be changed out of the data area of A.

The continuous data in the recorded flight data include engine high-pressure rotor speed, throttle location, back pod vibration value, exhaust temperature; metal contents in lubricating oil of the switching value exceed standard. These simulated data are expanded into 12 types of data, namely, the rotor speed, variation and change rate of two-stage compressor, the location, variation and change rate of throttle, the vibration value, variation and change rate of back pod, and the temperature, variation and change rate of the exhaust. Because the variation and change rate of the above-mentioned parameter is changed by using maximum–minimum standardization, its absolute value is not used, and it is normalized into interval [0, 1]. In order to ensure the normal operation of SOFM, the network weight must be normalized, and its formula is:

Initial value:

$$W_j(n+1) = \frac{W_j(n)}{\left\| W_j(n) \right\|} \tag{5.10}$$

Iterative equation:

$$W_j(n+1) = \frac{W_j(n) + \eta(n)[x(n) - W_j(n)]}{\left\| W_j(n) + \eta(n)[x(n) - W_j(n)] \right\|}, \quad j \in \Lambda_i(n) \tag{5.11}$$

Otherwise, $W_j(n+1) = W_j(n)$ in which $\Lambda_i(n)$ uses the rectangle network.

2. Confirmation of membership degree matrix

The data U of one sortie of flight is selected and the primary component features of the above-mentioned parameters are extracted. The procedure for extracting the primary component features is as follows.

Step 1. Estimate the autocorrelation matrix $\boldsymbol{R}_u$ of U according to k training samples of U;

Step 2. Solve the characteristic equation $|\boldsymbol{R}_u - \lambda| = 0$ of $\boldsymbol{R}_u$, and get the characteristic root $\lambda_1, \ldots, \lambda_n$;

Step 3. By using the above equation, the m characteristic roots corresponding to the principal component and the reduced dimension T_m of the data U of the corresponding sortie are calculated;

$$\sum_{i=1}^{m} \lambda_i / \sum_{i=1}^{n} \lambda_i > 85\% \quad (5.12)$$

Step 4. Get m components of reduced dimension T_m by using Eq. (5.5.12), and use them to form the transformation matrix;

$$\boldsymbol{\lambda}_{ia(i)} = a(i)\boldsymbol{R}_u \ (i = 1, 2, \ldots, m) \quad (5.13)$$

Step 5. Get principal features V by using Eq. (5.5.13), that is

$$V_m \times l = T_m \times nU_m \times l \quad (5.14)$$

Extract the primary component features of v', calculate $\boldsymbol{R}_u = E[\boldsymbol{xx}^T]$, $\boldsymbol{R}_u$ must be real symmetric matrix. Its characteristic value is nonnegative real number, and the vectors of the different corresponding characteristic values are orthogonal. After getting the characteristic value, according to the application experience, the components whose contribution rate is less than 5% are eliminated. Is should be ensured that the number of the dimension is less than 10 and that the contribution rate of single vector is more than 5%.

$$\frac{\lambda_i}{\sum_{i=1}^{n} \lambda_i} < 5\%$$

The actual input of the input vector T of SOFM neuron, that is

$$\boldsymbol{T} = \boldsymbol{AS}' \quad (5.15)$$

Input the real value according to the flight manual, the maximum–minimum standardization of vector S is

$$\boldsymbol{S} = \begin{bmatrix} \min A \\ \max A \end{bmatrix} = \begin{bmatrix} 10 & -68 & 5 & 0 & 0 & 0 & 0 & 0 & -5 & -40 & -20 & -25 \\ 110 & 45 & 100 & 1200 & 5 & 30 & 10 & 300 & 5 & 40 & 20 & 25 \end{bmatrix} \quad (5.16)$$

The input vector is S, and it is S' after extracting the principal feature. The input vector T is the input of the actual SOFM neuron. That is

$$\boldsymbol{T} = \begin{bmatrix} as'_{00} & as'_{01} & as'_{02} & as'_{03} & as'_{04} & as'_{05} & as'_{06} & as'_{07} & as'_{08} & as'_{09} \\ as'_{10} & as'_{11} & as'_{12} & as'_{13} & as'_{14} & as'_{15} & as'_{16} & as'_{17} & as'_{18} & as'_{19} \\ as'_{20} & as'_{21} & as'_{22} & as'_{23} & as'_{24} & as'_{25} & as'_{26} & as'_{27} & as'_{28} & as'_{29} \end{bmatrix} \quad (5.17)$$

The $\boldsymbol{A}$ matrix of Eq. (5.5.16) is from the size of the characteristic root after conducting principal feature analysis. Since the ratio reflects the contribute rate of components to whole variance, the membership degree matrix acquired from this contribution rate is:

$$\boldsymbol{A} = [\boldsymbol{A}_1, \boldsymbol{A}_2, \ldots, \boldsymbol{A}_{10}] \tag{5.18}$$

A matrix is the membership degree matrix which is got from the experts' experience.

$$\boldsymbol{A} = \begin{bmatrix} 2 & 1 & 5 & 10 & 15 & 10 & 7 & 20 & 12 & 4 \\ 2 & 1 & 5 & 25 & 20 & 24 & 10 & 25 & 14 & 15 \\ 2 & 1 & 5 & 30 & 25 & 30 & 30 & 30 & 15 & 30 \end{bmatrix} \tag{5.19}$$

Select the data from one sortie to be the input vector, and extract the principal features of the above-mentioned 12 parameters. Altogether 10 principal feature parameters are extracted, which are respectively two-stage compressor rotor speed, throttle location, back pod vibration speed, variation of the two-stage compressor rotor speed, variation of the throttle location, variation of exhaust temperature, change rate of the two-stage compressor rotor speed, change rate of the throttle location, change rate of the back pod vibration speed, and change rate of exhaust temperature.

3. The actual testing result of the flight data

The fault phenomenon of one engine malfunction is: No. 1 engine is tested by using the ground testing equipment, and it is found that the back pod vibration value is greater. After examination, it is found that the bearing is severely abraded, which results in the replacement of engine before the scheduled time. Now, the engine back pod vibration value of the flight parameter is checked, and it is found that this value is bigger. After the scheduled checkup of the engine, the flight parameter of the normal engine is the input value of group *A*, the flight parameter of the engine which has been malfunctioning for 4 h before being replaced is the input value of group *B*, and the data of the time from the metal signal starts to flash to the moment the aircraft lands is the input value of group *C*. The data in each group has 1200 frames, and the corresponding distribution graph is shown in Fig. 5.10a.

After adopting SOFM network to reflect the multidimensional data onto two-dimensional plane, the whole data displays in three clusters, but the distance between two of the clusters is comparatively close. Now the shrinking clustering method is used to conduct self-clustering, and the result is show in Fig. 5.11b. Table 5.1 is the simulated process parameters, and Table 5.2 is the corresponding expert interpretation. The result of the expert interpretation is identical to that of the self-clustering analysis. The difference between the states represented by the data in group B and C is comparatively obvious.

During airline flying, "dangerous vibration" signal light flashed while using this engine, but nothing was found while it was check on the ground. After replacing the electronic assembly, such fake warning never appeared, thus it belongs to the fake warning caused by the electronic assembly. Now this method is adopted to conduct

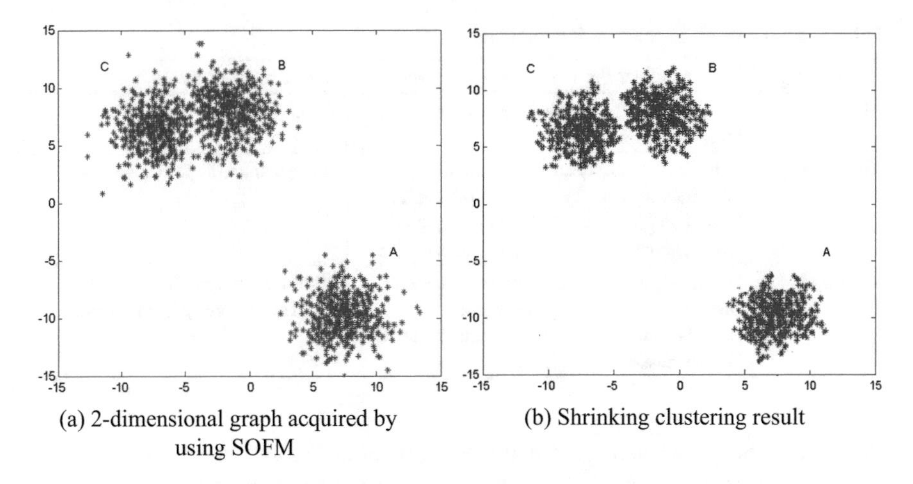

(a) 2-dimensional graph acquired by using SOFM

(b) Shrinking clustering result

Fig. 5.10 Comparison graph 1 of the result of shrinking clustering

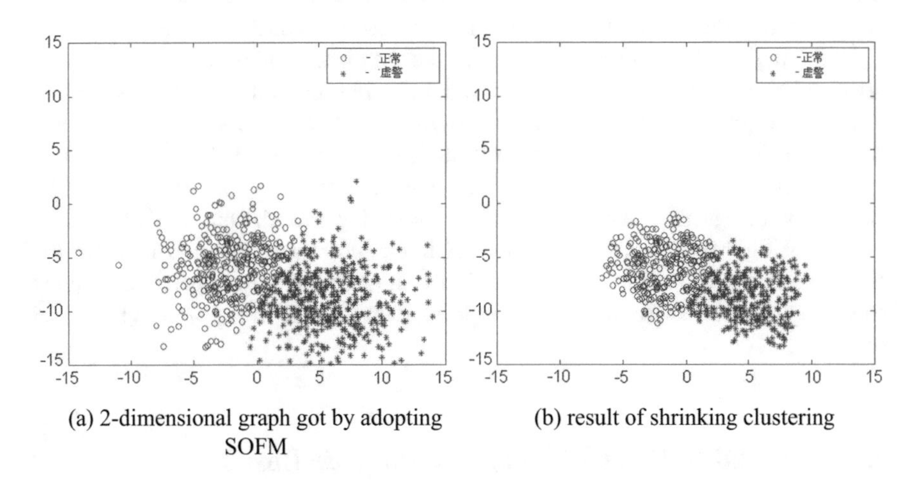

(a) 2-dimensional graph got by adopting SOFM

(b) result of shrinking clustering

Fig. 5.11 Comparison graph 2 of the result of shrinking clustering

Table 5.1 The simulated shrinking process parameters of flight data

	Before shrinking	After shrinking	Ratio
Number	1200	845	70.4%
Area	426 (self-defined distribution density)	321	75.3%
Scattered points are eliminated	–	355	–
Number of the ranking in ascending order	–	5	–
Tightness process value	0.4783 (ICD)	5.823 (ECD)	12.174

Table 5.2 The result of the data interpretation

Data group	The result of the parameters observed	Result of interpretation
Group A	The parameters correspond to the requirements	Good
Group B	The rotational speed automatically declines at level fight; about 40 s after No. 1 engine is at its full throttle state, the exhaust temperature continues to rise	Performance degrades obviously
Group C	Metal signal flashes to show that metal content in No. 1 engine exceeds standard; the back pod vibration value in air surpasses some value	Malfunction symptom, engine replaced before scheduled time

clustering analysis of the fight data in this period of time after the data are output by SOFM network. From Fig. 5.11, it can be seen that the result got from clustering analysis is basically the same as that when the engine is at its normal condition(o), which shows that the engine was at good condition when the fake warning(*) was given and that the shrinking clustering method itself is quite effective.

Searching for the mode represented by those unknown flight data is one of the main methods used to solve the expert system bottleneck problems. In this section, flexible-size grid high-dimensional data shrinking clustering method is provided. Among this method, integral movement and flexible-size grid method is a good way to conduct quick clustering of data, and the data close to one another can be effectively categorized when conducting clustering calculation; the method of measuring tightness by using ICD and ECD can compress data and measure the effectiveness of clustering. This method is quite good at multidimensional quick clustering, and is not limited to the type of data, thus it has a wide application scope.

5.2 Mutability Fault Diagnosis Arithmetic Based on Expert System

The fault of equipment is generally divided into mutability fault (hard fault) and graduality fault (soft fault). The symptoms of mutability fault are sudden loss of function or great change of output result. Any subsystem or airborne equipment of aircraft may have some kind of fault inevitably. How to diagnose the fault in time so as to minimize the loss is one of the main tasks of aircraft status monitoring. The performance parameters which reflect the key equipment of the aircraft are included in the flight data, which can be analyzed to monitor the status of the aircraft equipment in the following aspects: (1) reporting of single parameter over limit; (2) reporting of multi-parameter coordination out-of-tolerance; (3) recording of the working time of life-limiting components. This section mainly studies the method and procedure of diagnosing mutability fault based on the flight data of one sortie by using the theory of expert system.

5.2.1 Theory of Expert System

1. Basic concepts

There is not a strict definition for expert system so far. Expert system is generally perceived as "the computer system which can simulate human experts' ability to make decisions". Simply speaking, expert system contains the knowledge acquired from the experts in relevant fields, and the user can solve the problem by resorting to the knowledge contained in the expert system.

The earliest research work of expert system was first started by Feigenbaum, Lederberg, Shortliffe and Buchanan, etc., from Stanford University in late 1960s and early 1970s. Most of the expert systems before the mid-1970s mainly belonged to the type of explanation or the type of system which diagnosed faults and illnesses. The majority of the problems solved by them were ones which could be broken down. In the later part of 1970s, other types of expert system gradually appeared, such as designing, planning, controlling, etc. During this period of the time, the system of the expert system experienced some significant change. It had developed from the initial single knowledge database and single inference engine into multi-knowledge database and multi-inference engine, from centralized expert system into distributed expert system. With the re-rise of artificial neural network, people started to develop neural network expert system and the expert system which combined symbol processing with neural network. In recent years, with the development of the study of machine learning, people have gradually replaced the primary exact presentation and reasoning or comparatively simple inexact reasoning model, and have also conducted research into monotonic reasoning, inductive reasoning, etc.

Expert systems have the following commonness:

(1) It is a system with certain artificial intelligence;
(2) It contains the expert-level knowledge of the relevant fields;
(3) It can use AI technology to simulate the thinking and reasoning process of human experts in solving problems to solve the difficult problems in relevant fields and have the same result as experts.

In general, expert systems have the following characteristics:

(1) Instructiveness. Much of the knowledge in the expert system comes from the experiences of the experts in relevant fields. These experiences, when measured in a big scope, may not be complete or accurate, but in the relatively limited space, they are effective for a special issue. The expert system can apply the instructive searching strategy to compare, judge, identify and reason

the knowledge stored in the knowledge database. Therefore, the knowledge is instructive.

(2) Transparency. Expert system can explain its problem-solving process and basis. It can also provide answers to the questions related to its knowledge.
(3) Flexibility. According to the new requirements of experts and the experiences accumulated during the process of the expert system's own "work", it can modify and expand the knowledge continually so as to improve its problem-solving ability.
(4) Symbol operation. Expert system uses symbols to show knowledge, and uses symbol set to describe problems, and emphasizes the processing and calculation of symbols.
(5) Uncertainty reasoning. Most of the problems solved by expert systems are uncertain, and the knowledge used is based on experience, and the relevant information based on which the problems are solved is uncertain. Therefore, expert systems can conduct reasoning by combining vague and uncertain information and knowledge.

The above-mentioned characteristics of expert system are the key factors of expert system to be distinguished from any other type of program system, the standards to judge whether program system is expert system or not, and also the important aspect to determine whether the functions of an expert system are strong or weak.

2. Components and functions

From the aspect of the functions needed to be performed by expert system, the system must consist of two key components: the knowledge database to support the system to solve problems, and the knowledge process component which conducts operations of the knowledge database, including comparing, identifying, and reasoning of information. A basic expert system can by formed only if these two components are organically combined together. When forming such a system, not only the uniqueness of knowledge database and knowledge processing component should be considered, but also the coordination relationship between the two. Then, it is convenient to modify and expand the knowledge database without destructing the structure of the current system, thus building the expert system processing platform. For an expert system with comparatively complete functions, it does not consist of the two basic components alone (knowledge database and knowledge processing component). Instead, it is composed of a more complex software system. Currently, a representative expert system generally consists of the following components: the knowledge database to store knowledge; the inference engine to process knowledge; the database to store all the concept names, properties and intermediate information and data; the explaining program to explain the problem-solving process, and knowledge acquiring program and system organizing program to be used to build, modify and delete knowledge in the knowledge database. Its simple structure is shown in Fig. 5.12.

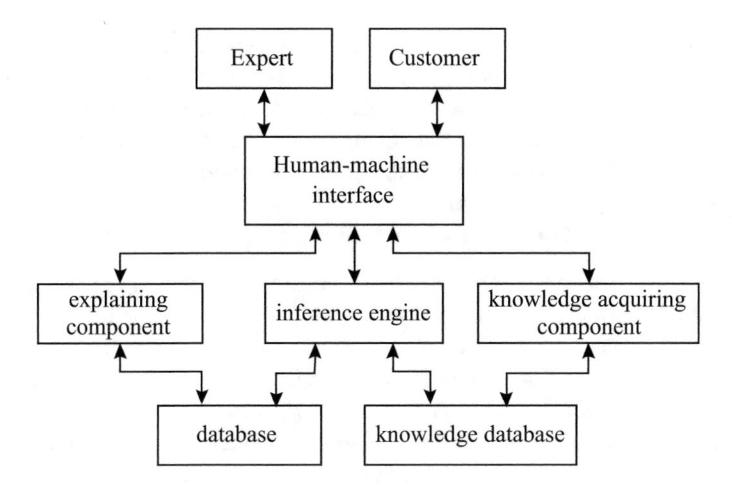

Fig. 5.12 The structure of expert system

(1) Inference engine. Inference engine is the "thinking" structure of the expert system, and the key part of the structure of the expert system. Its task is to simulate the thinking process of field expert, control and conduct the problem-solving process. Based on the known facts, it can conduct the reasoning work and get the answer to the questions according to certain reasoning methods and control strategy by using the knowledge stored in the knowledge database.
(2) Knowledge acquiring component. The basic task of knowledge acquiring component is to input knowledge into knowledge database, and to be responsible for maintaining the consistency and integrity of knowledge so as to build a database with good performance. What needs to be mentioned is that the knowledge acquiring functions and the ways to realize this are quite different for different expert systems.
(3) Knowledge database. Knowledge database is the component to store knowledge. It is used to store the theoretical knowledge, the experience knowledge of experts, and relevant facts in its own field. The knowledge in the knowledge database comes from knowledge acquiring component, and at the same time, knowledge database has to provide the inference engine with the knowledge needed to solve problems.
(4) Human machine interface. Human machine interface is one between the expert system and field experts or knowledge engineer and general user, which consists of a set of software and relevant hardware, used to finish the input and output work. Through this interface, field expert or knowledge engineer inputs knowledge so as to update and perfect knowledge database, general user inputs the problems to be solved, the known facts, and the problems proposed, and the system outputs the calculated result, answers user's questions, or seeks further facts.

(5) Database. Database is the working memorizer which is used to store the initial facts provided by the user, problem description and intermediate result, final result and working information of the system functioning process.
(6) Explaining component. The explaining component can explain the behavior of the expert system, and is one of the major characteristics to distinguish expert system form general system.

3. General information of the research of expert system

Currently, the theory of expert system is becoming increasingly perfect, and has been applied extensively. In recent years, with the appearance of expert system developing software, more medium and small expert systems are entering into human life and production. There is no exception in the field of aviation maintenance. In the 7th ICRMS held in France, expert system was considered to be one of the three developmental directions of reliability maintenance. Swiss DC-9 aircraft diagnosis expert system and Canadian large maintenance planning expert system of aircraft have already put into actual operation. In China, many civilian aircraft maintenance and fault diagnosis expert systems are under research and development. In addition, many special personnel from three Aviation and Aeronautic universities and relevant research institutions are conducting researching concerning expert system. However, the research objects of the published papers concerning expert system are basically limited to large civilian aircraft. Of course, there are still many problems which need to be solved during the process of the research and development of expert system, such as the issue of the completeness of knowledge, the automatic acquisition of knowledge, the presentation and employment of deep-level knowledge, the handling of distributed knowledge, the cooperation and integration of multi experts, the reasoning of common sense knowledge, etc. These issues need to be further studied, and it also needs the development of other field such as AI.

Expert system converts people's long-term practical experience and abundant fault information knowledge into rules which can be used by computers, then these rules are provided to the knowledge database, and the real-time data of the diagnosed system are provided to the database as well. Expert system conducts reasoning and analysis to the knowledge and data stored in the database by using the rules of the knowledge database, and then it can find the final faults or the possible faults. The design of fault diagnosis system by adopting the expert system accords with the requirements of the inspection and reality of the aviation facilities. Many years of maintenance experience of the experts and artificial intelligence level of special fault handling cannot be replaced by general methods and algorithm. Therefore, expert system becomes the first choice to be used to design and develop the system related to flight data processing. In the flight data processing system, the information recorded by the flight data recording system is processed by identifying reference engine according to the knowledge in the knowledge database, and then many different methods are provided to handle the user's search. Because aircraft faults is related to many indirectly recorded parameters, users are allowed to modify the knowledge database through the simple interface by themselves so as to ensure the reliability of the knowledge database.

5.2.2 Main Functions of the Aircraft Fault Diagnosing Expert System

Based on the maintenance experience of aircraft maintenance experts for the different systems of a given type of aircraft, the historical data are processed to extract the important parameters from the flight data, to establish the performance analysis charts of the different systems of the aircraft, and to provide support for maintenance and decision. For example, after calculating the comprehensive parameters of the engine, the display surface of the calculated parameters curve is shown in Fig. 5.13. This figure shows the result of the calculation by adopting fixed weight value synthesis parameter algorithm. The abscissa shows the number of sortie, and the ordinate shows the scope of weight value [0, 1].

The samples are the data of the recent 1100 sorties of one engine taken from the FDR. The samples are taken when the engine is at the stable state, which means

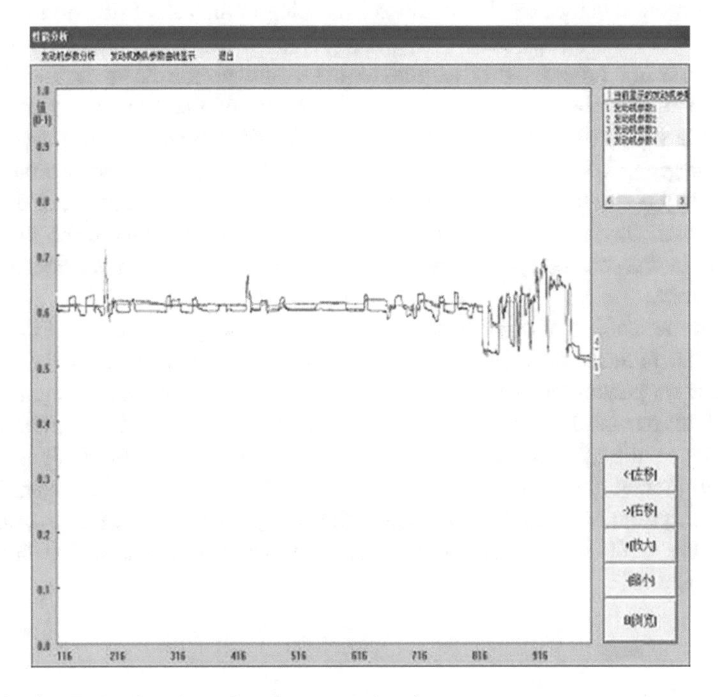

Fig. 5.13 The display interface of engine comprehensive parameter curve

both the engine throttle and engine low-pressure rotor speed are pointing at the stable marker. The condition for engine throttle to be at stable state is that the changing rate of engine throttle is less than 1.5(°)/s for more than 10 s, and at the same time, the signal for "engine vibration limit" is not "1". The stable state for engine low-pressure rotor speed is that the change rate of N is less than 0.95%/s for more than 10 s. The calculating formula for the comprehensive parameter of fixed weight value is shown below:

$$\left.\begin{array}{l} Q(t) = \frac{\sum_{s=1}^{k} v_s \widehat{x}_s(k)}{\sum_{s=1}^{k} v_s} \\ k = 1, 2, \ldots, m \end{array}\right\} \tag{5.20}$$

In this formula, m is the number of the parameters of the engine which are monitored; v_s is the weight of parameter $x_s(k)$, which shows the importance of this parameter while estimating the monitored status. The weight value of different parameters can be got from analyzing the information of each parameter, or according to the relevant experience. The limitation for v_s is:

$$\sum_{s=1}^{m} v_s = 1 \quad 0 \leq v_s \leq 1; \tag{5.21}$$

These weight values can be modified according to the actual situation while they are used so as to acquire more reasonable comprehensive parameter value. In the real monitoring process, the method to judge the performance is the changing condition of the observed value. The performance of the engine is perceived as decreasing obviously when the value is continually vibrating in a comparatively greater degree. This shows that the engine needs to be checked thoroughly. In Fig. 5.13, the calculated parameter started to vibrate continually near the 800th sortie, which accorded with the time limit for this type of engine to be thoroughly checked in the workshop, and with the real situation of obvious decreasing of performance.

When the reading of flight data shows a fault, it can give enough information to remind the fault according to the real maintenance manual to help the frontline maintenance personnel locate and solve the problem. It can also search the historical fault database on the basis of the display list and display the searching result according to the related items, so as to ensure sufficiency of the fault reminding information and the reliability of the source. The result of the searching for hydraulic system faults in the historical fault database is shown in Fig. 5.14 in excel format. The faults in this figure is automatically listed according to the date of the faults happened.

1	飞机号	发现日期	故障件名称	故障件型别	故障时机及现象	部位划分	故障专业	后果	判明方法
2		1995-2-21	液压油泵		液压油渗漏		机械	未影响	
3		1995-8-11	液压油泵		分液活门卡死，轴断		机械		
4		1996-3-5	液压泵		液压一系统最低油量信号牌亮	发动机舱	机械	其它	使用
5		1996-3-5	液压泵		液压一系统最低油量信号牌亮	发动机舱	机械	其它	使用
6		1996-12-3	液压油泵		泵出口无压力		机械		
7		1996-12-3	液压油泵		油量少指示灯亮		机械	未影响	
8		1997-7-7	液压油泵		油量少指示灯亮		机械	未影响	
9		1997-9-6	液压油泵		飞行下降，油泵工作压力底		机械		
10		1998-6-3	液压油泵		不供压		机械	未影响	
11		1998-6-6	液压油泵		工作时金属含量过多	发动机	机械	未影响	试车
12		1998-8-19	液压油泵		工作无压力		机械		
13		1998-12-1	液压油泵		工作信号灯不亮		机械		
14		1999-3-25	液压泵		滤网上有大量金属屑		机械		
15		1999-6-8	液压油泵		压力检查按钮灯不亮		机械		

Fig. 5.14 The auxiliary function of historical fault information

5.2.3 *Implementation and Evaluation*

The main contents for evaluation are as follows: whether the result of the handling of the flight data with real fault is accurate; how the inference engine operates after one rule in the knowledge database has been changed; and whether the system can automatically store the result of the handling of historical data.

The ways for examination are: compare the result got from computer processing with that of human analysis; check whether the output form of each flight event is in accordance with the requirement of the regulation; check whether the time scale of the event is accurate, including the field experts' analysis of the same flight data through data listing display and parameter curve. The evaluation of it accuracy and generality is provided below.

1. The evaluation of the accuracy of the result of flight data handling

The flight data of a certain period of time are 926 mega-frames, with 1852s total recording time. Then the data is analyzed by using knowledge database, and the result is shown in Fig. 5.15, among which item 38 is a dangerous vibration event. Now the frame where the raw data lies is analyzed in detail.

Figure 5.16 shows that the switching value of the dangerous vibration appeared at No. 349 mega-frame. After calculation, it is confirmed that the corresponding astronomical time of this frame is 8:29, which is exactly the same as the result of manual data process, thus showing that the data analysis of expert system is accurate and reliable.

判据管理 数据分析 帮助 退出数据分析

号码	判据名称	指定数据段内共发生次数	第一次发生时间
27	机翼失火	0	
28	BCY舱失火	0	
29	发动机短舱失火	0	
30	一发停车手柄放"停车"位	8	7 : 55
31	二发停车手柄放"停车"位	4	8 : 20
32	三发停车手柄放"停车"位	4	8 : 20
33	四发停车手柄放"停车"位	4	8 : 20
34	一发接通反推	1	10 : 03
35	二发接通反推	3	8 : 30
36	三发接通反推	3	8 : 30
37	四发接通反推	3	8 : 30
38	一发危险振动	2	8 : 29
39	二发危险振动	0	
40	三发危险振动	0	
41	四发危险振动	0	

Fig. 5.15 The result of the analysis to a certain segment of flight data

履历显示 选择要查看的参数 选定参数显示 曲线图操作 帮助 退出参数显示

号码	大...	小...	飞行无线电高度	飞行气压高度	真空速	磁航向	倾斜角	一发危险振动	二发...	三发...	四发...
3492	349	3	0.00	176.47	79.31	321.82	0.56	1	0	0	0
3493	349	4	0.00	176.47	79.31	321.82	0.56	0	0	0	0
3494	349	5	0.00	176.47	79.31	321.82	0.56	0	0	0	0
3495	349	6	0.00	176.47	79.31	321.82	0.56	0	0	0	0
3496	349	7	0.00	176.47	79.31	321.82	0.56	1	0	0	0
3497	349	8	0.00	176.47	79.31	321.82	0.56	1	0	0	0
3498	349	9	0.00	176.47	79.31	321.82	0.56	1	0	0	0
3499	349	10	0.00	176.47	79.31	321.82	0.56	1	0	0	0
3500	350	1	0.00	176.47	79.31	321.82	0.56	1	0	0	0
3501	350	2	0.00	176.47	79.31	321.82	0.56	1	0	0	0
3502	350	3	0.00	176.47	79.31	321.82	0.56	1	0	0	0
3503	350	4	0.00	176.47	79.31	321.82	0.56	1	0	0	0
3504	350	5	0.00	176.47	79.31	321.82	0.56	1	0	0	0
3505	350	6	0.00	176.47	79.31	321.82	0.56	0	0	0	0
3506	350	7	0.00	176.47	79.31	321.82	0.56	1	0	0	0
3507	350	8	0.00	176.47	79.31	321.82	0.56	1	0	0	0
3508	350	9	0.00	176.47	79.31	321.82	0.56	0	0	0	0
3509	350	10	0.00	176.47	79.31	321.82	0.56	1	0	0	0
3510	351	1	0.00	176.47	79.31	321.82	0.56	1	0	0	0
3511	351	2	0.00	176.47	79.31	321.82	0.56	0	0	0	0

Fig. 5.16 The real display of a certain segment of data

2. The evaluation of generality when date frame and aircraft type changes

First, as for three formats of data and two types of databases which are recorded in the flight data system of this type of aircraft, the designed expert system can read in and handle the data effectively. Second, the expert system can choose by itself the frame header, space mark definition, and frame length, which shows that it has common data reading function. Thirdly, both the self-defined database file and the database defined in WINDOWS platform can read in the data. It can read and handle the data on the imported large commercial aircraft and cargo aircraft equipped with this type of flight data recording system.

To sum up, the aircraft equipment fault diagnosis expert system based on flight data has the following characteristics:

(1) The knowledge database is completely separated from inference engine;
(2) The knowledge database adopts the productive regulation description, which is open to customer and easy to maintain;
(3) The computing symbols defined by the inference engine compiling system and the output format of event can satisfy customer's requirements for monitoring the aircraft equipment status;
(4) Inference engine adopts the positive accurate reasoning technique of data drive strategy and regards the end of the flight data as the mark of the end of reasoning, thus the phenomenon of "combination explosion" will not appear.
(5) The explanation component of the system is comparatively perfect.

5.3 Gradual Fault Diagnosis Arithmetic Based on Dynamic Principle Component Analysis

As for the faults in equipments, gradual fault has a phenomenon that the result of the measurement will shift, and the shift becomes more severe with the passage of time. The arithmetic for fault diagnosis include fault diagnosis based on physical model, fault diagnosis based on mathematical model, and fault diagnosis based on intelligent way. The application of the arithmetic to paroxysmal faults is proven to be of good engineering effect. While for gradual faults, the general arithmetic is to extract the characteristic parameters of the process data, change these parameters into paroxysmal faults, and then handle them, which results in the low diagnosis accuracy and low sensitivity. There is a strong connectivity among multivariable flight data, which provides analytic redundancy for solving gradual fault diagnosis problem. Traditional principle component analysis only deals with the observed data of a given time, and perceives that the statuses of the observed object at different times are isolated and unrelated (stationary). In reality, however, many characteristic data of physical object are continued in time (dynamic). This part provides the gradual fault diagnosis arithmetic based on dynamic principle component analysis.

5.3.1 Overview of Principle Component Analysis

Principle Component Analysis (PCA) is a multivariate statistical method, which can be used to analyze the measured data with noise that are related to height. The

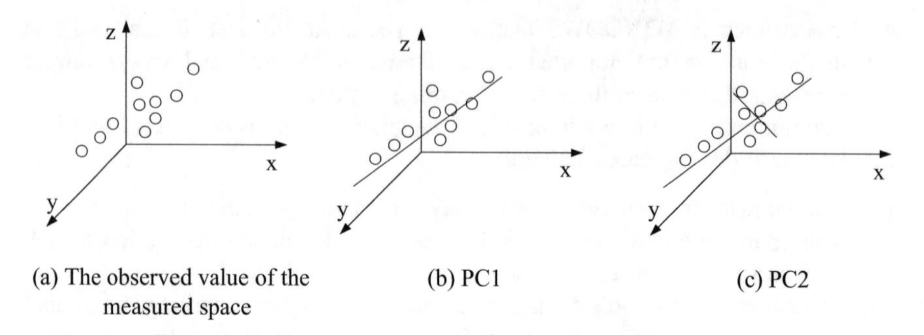

(a) The observed value of the measured space (b) PC1 (c) PC2

Fig. 5.17 Geometrical explanation of PCA

method used is to project the high-dimensional information onto low-dimensional space, and preserves the main process information. PCA, first proposed by Pearson in 1901, is a multivariate statistical analytical skill which changes many interrelated affecting factors of a process into a few independent factors, and it is categorized into standard multivariate statistical method. In 1933, Hotelling made some improvements to this method, which is the basis for different types of the current principle component analysis.

PCA is mainly used to describe the data changing situation of single data matrix. PCA first calculates a vector, called PC1, to describe the biggest changing direction of the data. PC2 is orthonormal with PC1, describing the biggest remaining changing direction. In three-dimensional space, one geometrical explanation of PCA is shown in Fig. 5.17. In this figure, most of the data are distributed in one plane. Therefore, in most cases, generally, a few principle components can be used to describe the high-dimensional data matrix.

Suppose $\boldsymbol{X}$ is the data matrix of $n \times m$. Its rows correspond to the samples, and its columns correspond to the variables. Matrix $\boldsymbol{X}$ can be decomposed into:

$$\boldsymbol{X} = \boldsymbol{TP}^T = t_1 p_1^T + t_2 p_2^T + \cdots + t_m p_m^T \tag{5.22}$$

where $\boldsymbol{T} = [t_1, t_2, \ldots, t_n]$ is called score matrix; $t_i \in R^n$, $i = 1, 2, \ldots, n$ is called score vector, and is also called the principle component of $\boldsymbol{X}$; $\boldsymbol{P} = [p_1, p_2, \ldots, p_m]$ is the load matrix; $p_i \in R^m$ is the load vector; $i = 1, 2, \ldots, m$.

The score vector is orthonormal with load vector. Multiply p_i to both sides of formula 5.22, we have:

$$\boldsymbol{X} p_i = t_1 p_1^T p_i + t_2 p_2^T p_i + \ldots + t_m p_m^T p_i = t_i$$

It can be seen that the projection of data matrix $\boldsymbol{X}$ on load vector p_i is the corresponding score vector t_i, whose length $\|t_i\|$ reflects the covering degree of data matrix on p_i direction. When there is a certain degree of linear correlation among the variables of the data matrix, the change of the data matrix will be mainly on the direction of k load vectors which are at the front. The projection of the load vectors at the back will be very few (mainly caused by the measurement noise and system error). Then, data matrix $\boldsymbol{X}$ can be decomposed into:

$$\boldsymbol{X} = t_1 p_1^T + t_2 p_2^T + \cdots + t_k p_k^T + \boldsymbol{E} = \boldsymbol{T}^T \boldsymbol{P} + \widetilde{\boldsymbol{T}}\widetilde{\boldsymbol{P}}^T$$

In this formula, $\boldsymbol{P} = [p_1, p_2, \ldots, p_k] \in \boldsymbol{R}^{m \times k}$ is the principle component load matrix; $\tilde{\boldsymbol{P}} = [p_{k+1}, p_{k+2}, \ldots, p_n] \in \boldsymbol{R}^{m \times (n-k)}$ is the residual error load matrix; $\boldsymbol{T} = [t_1,\ t_2, \ldots, t_k] \in \boldsymbol{R}^{n \times k}$ is the principle component score matrix; $t_1, t_2, \ldots, t_k$ and $\|t_1\| \geq \|t_2\| \geq \cdots \geq \|t_k\|$ is principle component load vector; $\tilde{\boldsymbol{T}} = [t_{k+1}, t_{k+2}, \ldots, t_n] \in \boldsymbol{R}^{n \times (m-k)}$ is the residual error score matrix.

In the formula, $[\boldsymbol{P}, \tilde{\boldsymbol{P}}]$ is unit orthogonal matrix; $[\boldsymbol{T}, \tilde{\boldsymbol{T}}]$ is orthogonal matrix; $\boldsymbol{TP}^T$, called principle component space, is the projection of $\boldsymbol{X}$ on the direction of principle component load vector; $\boldsymbol{E} = \widetilde{\boldsymbol{T}}\widetilde{\boldsymbol{P}}^T$, called residual error, represents the projection of $\boldsymbol{X}$ on the direction of load vectors other than the principle component. Because error matrix $\boldsymbol{E}$ is mainly caused by measurement noise and system error, removing $\boldsymbol{E}$ equals to getting rid of the noise interference, which will not cause the loss of effective information. Thus, $\boldsymbol{X}$ can be approximately shown as:

$$\boldsymbol{X} \approx t_1 p_1^T + t_2 p_2^T + \cdots + t_k p_k^T = \boldsymbol{TP}^T \tag{5.23}$$

It can be proven that the load vector of matrix $\boldsymbol{X}$ is actually the characteristic vector λ_i of the covariance matrix $\boldsymbol{X}^T\boldsymbol{X}$, and the principle analysis for $\boldsymbol{X}$ is actually the analysis for λ_i.

5.3.2 Dynamic Principle Component Analysis Method

Traditional PCA only deals with the observed data of a given time, and perceives that the statuses of the observed object at different times are isolated and unrelated (stationary). In reality, however, many characteristic data of physical object are continued in time (dynamic), such as the measured data of the engine sensors. In order to make the PCA method for stationary multivariate suitable for the analysis of dynamic multivariate process, the observed data of the previous time t should be added to the data matrix to establish a multivariate multi-time autoregressive statistical model, which is called Dynamic Principle Component Analysis (DPCA). The observed data matrix with added data is a three-dimensional matrix $\overline{\overline{\boldsymbol{X}}}_{n \times m \times t}$ with a $n \times m \times t$ (sample $\times$ variable $\times$ time), whose expression formula of time section is:

$$\boldsymbol{X}(i) = \begin{bmatrix} x_{11i} & x_{12i} & \cdots & x_{1mi} \\ x_{21i} & x_{22i} & \cdots & x_{2mi} \\ \vdots & \vdots & \ddots & \vdots \\ x_{n1i} & x_{n2i} & \cdots & x_{nmi} \end{bmatrix}_{n\times m} \quad i = 1, \cdots, t. \tag{5.24}$$

$\boldsymbol{X}(i)$ is the input data matrix of previous time $i - 1$.

5.3.3 *Fault Diagnosis Arithmetic Based on Dynamic Principle Component Analysis*

1. Diagnosis principle

Fault diagnosis arithmetic based on dynamic principle component analysis includes two sections:

(1) Establish the dynamic principle component model of the sample;
(2) Use fault indicator to trace the fault so as to realize the purpose of diagnosis.

The basic idea is to establish the dynamic principle component model of the sample first, and then to monitor the relativity among the characteristic quantity by using dynamic fault indicator. When the relativity among the characteristic quantity is broken, the fault indicator can trace this phenomenon so as to realize the purpose of fault diagnosis.

2. Dynamic principle component model

The dynamic principle component matrix arithmetic for any given time is:

Step 1: Initializing data. The observed data of different samples for t numbers of adjacent time are chosen as the initial dynamic observed matrix.

Step 2: Normalizing data deviation. From every column of the element of the data matrix $X(i)$ of each time, subtract the average value of the column, then divide it by the standard deviation of this column, and the deviation matrix $\overline{\overline{E}}$ is acquired.

Step 3: Calculating the first principle component. Choose any one column of $\overline{\overline{E}}$ as $\vec{t}$:

① $\boldsymbol{P} = \overline{\overline{\boldsymbol{E}}}^T \vec{t}$
② $\bar{\boldsymbol{P}} = \bar{\boldsymbol{P}}/\|\bar{\boldsymbol{P}}\|$
③ $\vec{t} = \overline{\overline{\boldsymbol{E}}}\bar{\boldsymbol{P}}$. if $\vec{t}$ converges, then go to ④, otherwise return to ①
④ $\overline{\overline{\boldsymbol{E}}} = \overline{\overline{\boldsymbol{E}}} - \vec{t} \otimes \bar{\boldsymbol{P}}$

Step 4: For the other columns in $\overline{\overline{E}}$, calculate all the principle components according to the following formula.

$$\begin{aligned}
&\overline{\overline{E}}^T(n, m, t) = \overline{\overline{E}}(m, n, t) \\
&P = \overline{\overline{E}}^T \rightharpoonup t \Rightarrow P(t, n) = \sum_{m=1}^{M} \overline{\overline{E}}(m, n, t) \rightharpoonup t(m) \\
&\|P\| = \sqrt{\sum_{t=1}^{T}\sum_{n=1}^{N} P(t, n)^2} \\
&\vec{t} = \overline{\overline{E}}\bar{P} \Rightarrow \vec{t}(m) = \sum_{t=1}^{T}\sum_{n=1}^{N} \overline{\overline{E}}(m, n, t)\bar{P}(t, n) \\
&\vec{t} \otimes \bar{P} = \vec{t}(m)\bar{P}(n, t)
\end{aligned} \tag{5.25}$$

During calculation, while new observed data are added, the frontmost observed data should be deleted at the same time so as to ensure that the number of the data which are analyzed is t. After all the principle components are calculated, dynamic fault indicators can be used to conduct analysis and diagnosis.

3. Dynamic fault indicator

PCA usually adopts Squared Prediction Error (SPE) to monitor the correlation of each principle component characteristics. SPE is usually described as the following formula:

$$\text{SPE} = e^T e = \|e\|^2 = x^T(I - PP^T)x \tag{5.26}$$

As for the judging regulations for PCA residual error, when the following formula is established, it can be judged that the correlation among variables is broken, which is also called fault indicator based on SPE.

$$\text{SPE} \geq cl(\alpha) \tag{5.27}$$

In this formula, $cl(\alpha)$ is the confidence limit of Gaussian Distribution.

The fault indicator based on SPE only adopts the single measured value to calculate so as to monitor the correlation of the current time. In order to make SPE to be more sensitive to the change of the observed value, the dynamic characteristics of the observed value can be chosen, combining with SPE, to construct new fault indicator.

Considering the consistency between the changing rate in the characteristic value dynamic attribute and the residual error changing rate, the changing rate and changing direction of the characteristic value are chosen as the new dynamic characteristics index to construct the Dynamic Squared Prediction Error (DSPE) fault indication, which is described in the following formula:

$$\begin{aligned}
&\text{DSPE} = \lambda \cdot x^T(I - PP^T)x \\
&\lambda = 1 + hv
\end{aligned} \tag{5.28}$$

In this formula, λ is the dynamic factor; h is the changing rate of principle component characteristics at current time; v is the changing direction of principle component characteristics at current time. When the value of DSPE is bigger than the given cl, it can be judged that there is fault.

DSPE fault indicator considers not only the correlation of the measured value at current time, but also the changing trend of the measured value, thus improving the sensitivity of the fault indicator.

5.3.4 Simulation Evaluation of Fault Diagnosis

Since the diagnosis object and application occasion are different, simulation diagnosis is conducted for flight data acquisition sensor and airborne equipment, respectively. The effects of the following three kinds of diagnosis arithmetic are compared: arithmetic 1, PCA diagnosis method based on SPE; arithmetic 2, DPCA diagnosis method based on SPE; and arithmetic 3, DPCA diagnosis method based on DSPE.

1. Sensor fault diagnosis

In actual fault diagnosis, apart from the fault of the airborne equipment abnormal, another type of fault appears because the acquisition sensor fault causes the characteristics parameter to be abnormal. If this kind of abnormal characteristics parameter is not analyzed, equipment fault may be incorrectly reported. Therefore, sensor fault should be diagnosed first.

The mechanism for sensor fault diagnosis is different from that for equipment fault diagnosis. The latter is to monitor the samples of multiple measured values of process performance parameters in the same period of time, while the sensor has only one measured value as the self-diagnosis sample to reflect the physical parameter in a given period of time. Therefore, the following approximation processing should be done while sensor fault is diagnosed: within several measuring cycles which are very short in time, the output of the measurement of sensor is regarded as the multiple measured samples for the same period of time.

Four types of typical faults which are very common for sensors are precision declining, deviation, invalid, and shifting. Four groups of samples, 500 samples for each group, are simulated and acquired. The frequency for sample acquisition is 1 time per second. The first 400 samples of each group are chosen as the training sample to construct DPCA model, and the rest of the samples are used for model testing, and fault data are introduced in the 400th sample. Figure 5.18 is the DSPE monitoring figure of DPCA under the fault of precision declining; Fig. 5.19 is the DSPC monitoring figure of DPCA under the deviation fault; Fig. 5.20 is the DSPC monitoring figure of DPCA under complete invalid fault; and Fig. 5.21 is the DSPC monitoring figure of DPCA under shifting fault. Set confidence limit $cl = 0.23$.

Compared with arithmetic 1, Table 5.3 provides the monitoring steps of fault diagnosis for using two types of arithmetic.

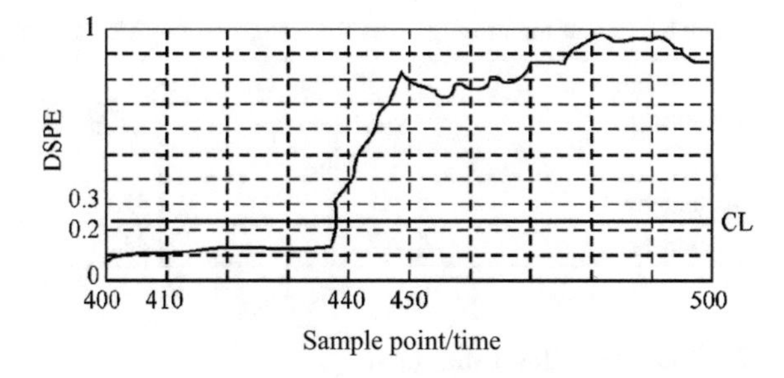

Fig. 5.18 DPCA monitoring under the fault of precision declining

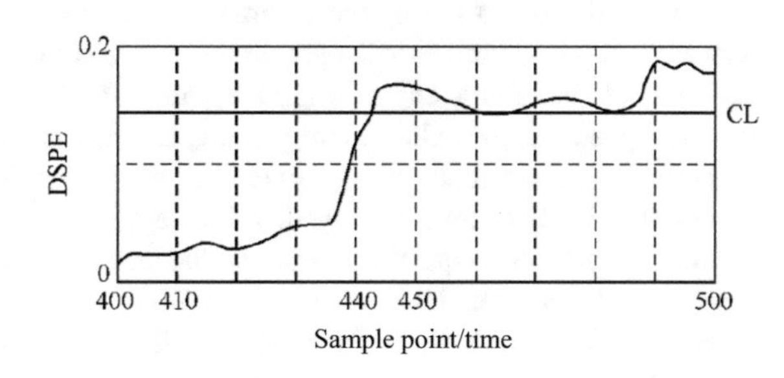

Fig. 5.19 DPCA monitoring under the deviation fault

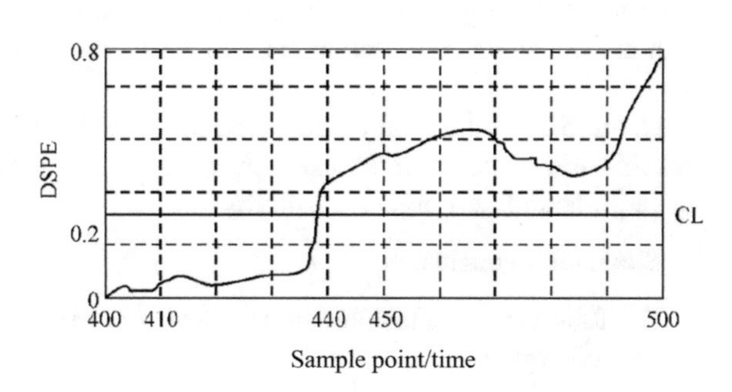

Fig. 5.20 DPCA monitoring under complete invalid fault

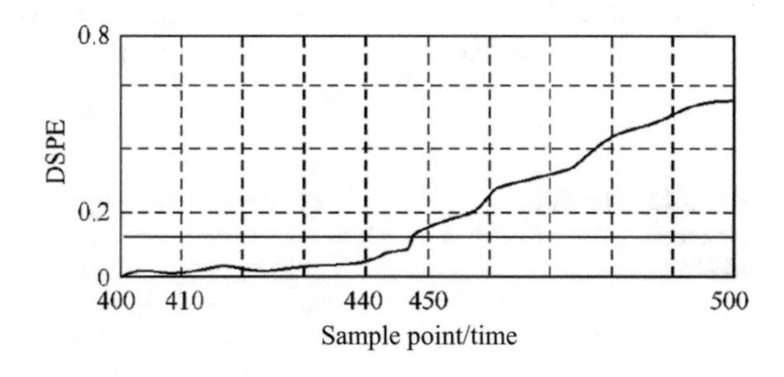

Fig. 5.21 DPCA monitoring under shifting fault

Table 5.3 Fault monitoring steps for using two types of arithmetic

Fault diagnosis arithmetic	Fault monitoring steps			
	Precision declining	Deviation	Complete invalid	Shifting
Arithmetic 1	451	450	446	462
Arithmetic 3	438	442	438	447

2. Equipment fault diagnosis

It is simulated that during one flying mission, the fault of engine rotor overspeed appeared. Seven characteristics are chosen to construct dynamic principle component model, including engine high-pressure rotating speed, engine low-pressure rotating speed, engine exhaust temperature, engine overspeed warning, engine noise spectral flow, short-time average over zero rate of engine noise, and engine noise spectral centroid. There are altogether 480 samples, with a frequency of 1 time per second for sample acquisition, and engine rotor overspeed is simulated at 80th step. Confidence limit 1 time per second $\boldsymbol{CL} = 21$.

Figure 5.22 is the fault diagnosis result of SPE value of arithmetic 1 and arithmetic 2. It can be seen from this Figure. that, after fault is introduced, SPE starts to increase greatly. Arithmetic 2 accurately monitors the happening of the fault at 90th step, while arithmetic 1 monitors the happening of the fault at 115th step.

Figure 5.23 is the fault diagnosis result of DSPE of arithmetic 3. It can be seen that this arithmetic monitors the happening of fault at 85th step, which shows that this arithmetic has a higher sensitivity.

3. Simulation conclusion

The following conclusion can be drawn from the fault diagnosis simulation of sensor and equipment:

(1) Because the dynamic nature of time is considered, when the same fault indicator (SPE) is used, dynamic principle component arithmetic has a higher sensitivity of diagnosis than principle component arithmetic.

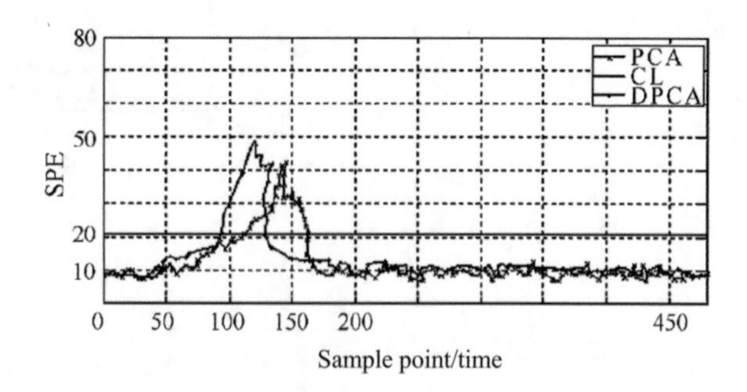

Fig. 5.22 The SPE monitored value of PCA and DPCA

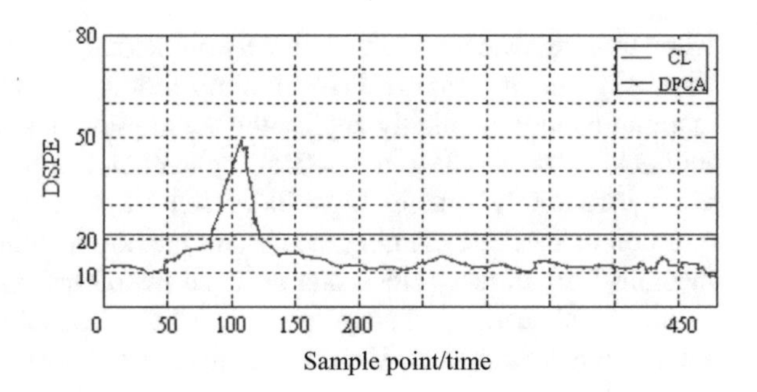

Fig. 5.23 The DSPE monitoring value of DPCA

(2) Because the changing trend of the principle component characteristics are considered, when the same fault diagnosis model (DPCA) is used, dynamic fault indicator has higher diagnosis sensitivity than traditional fault indicator.
(3) The fault diagnosis arithmetic which uses dynamic fault indicator and is based on dynamic principle component analysis is correct and effective. It also greatly improves the real-time performance of gradual fault diagnosis.

5.4 Engine Performance Parameter Prediction Based on WLS-SVM

This section deals with the application of WLS-SVM in time series prediction. Besides, the general framework of one-dimensional time series modeling prediction is also introduced. BIC standards are applied to select Embedding dimensions, and statistic-based model performance evaluation methods are provided. Modeling and prediction for typical engine performance parameters are based on WLS-SVM and AR models, respectively, and detailed comparative result is given. Experiments show that WLS-SVM has excellent extending ability as it adopts the new structural risk minimization principle, and its predictable range is wider and more accurate. As a result, as the monitor of aircraft performance parameters, WLS-SVM has a bright prospect of deployment.

5.4.1 Basic Theory of SVM

Data-based machine learning is an important research direction in the field of modern intellectual technology. The ultimate objective of machine learning is how to explore rules through observing data, and make further use of these rules to predict. In essence, traditional machine learning method is mainly based on empirical risk minimization principle. Empirical risk refers to the risk in the training sample set, and is normally expressed by mean square error (MSE). Theoretically, when training data tend to be infinitely numerous, the empirical risk converges to a true risk, thus the empirical risk minimization principle of the traditional machine learning method implicitly applies the assumption that the training sample data are infinitely numerous. But in practical application, this prerequisite is not always met, as is particularly true in high-dimensional space. And it is actually the basic question of machine learning theory and method including function approximation theorem, pattern recognition and artificial neural network.

Statistical learning theory specially studies machine learning principle under small sample conditions. Vapnik and others have been devoted to this theory since the 1960s. By the mid 1990s, statistical learning theory has drawn more and more attention, along with its development and maturity and also due to no substantive theoretical progress in artificial neural network and other learning methods. Based on this theory, a new general learning method—SVM was developed. Support Vector Machine (SVM) not only has many particular advantages in solving such problems as small sample, nonlinearity and high-dimensional pattern recognition, but also can be used in some other machine learning problems like function fitting. Besides, SVM successfully solves problems of high-dimension and local extreme value. Precisely because of the aforementioned advantages, SVM theory has become a hot research topic of machine learning after the artificial neural network, and will continuously promote the significant development of machine learning theory and technology.

SVM is the newest but the most practical part in statistical learning theory. The core concept of SVM was developed from 1992 to 1995, and it is still in the stage of developing. The theoretical foundation of SVM is statistical learning theory. SVM follows the structural risk minimization principle, and is mainly applied to solve problems of classification and regression in machine learning. Because SVM is developed on the basis of structural risk minimization principle, errors of sample points are minimized, and upper bound of model prediction errors is reduced simultaneously. Consequently, the model's generalization capability is enhanced. All these characteristics make SVM an excellent algorithm and a hot topic in machine learning study.

1. Mathematical description of regression problem

SVM method can effectively solve function fitting problem. Regression and classification have similar mathematical representations which differ mainly in the value of variable y. In the case of classification, variable y is either -1 or $+1$. But in regression, variable y can be an arbitrary real value. Regression mathematical representation is:

The given training sample set

$$\boldsymbol{T} = \{(x_1, y_1), (x_2, y_2), \ldots, (x_n, y_n)\} \in (\boldsymbol{X} \times \boldsymbol{Y})^n \tag{5.29}$$

In the formula: $x_i \in R^n$, $y_i \in R$, $i = 1, \ldots, n$. Suppose the training sample set is an IID (independent identically distributed) sample point which is selected according to a certain probability distribution $P(x, y)$ of $\boldsymbol{X} \times \boldsymbol{Y}$, meanwhile the loss function $c(x, y, f)$ is given, the regression problem is to find a function to make the expected risk $R(a) = \int c(c, y, a)\mathrm{d}P(x, y)$ to reach to its minimum. Probability distribution $P(x, y)$ is unknown, and the training set is known.

2. Loss function

When data noise is unknown, regression result is influenced by loss function. In 1964, Huber proposed a theory that helped to find the best strategy of selecting loss function when only the general information of noise model is known. However, this loss function proposed by Hubert is not suitable to SV (support vectors), whereas sparseness of solution is very important to massive data processing in a high-dimension space. In order to construct SVM for real-valued function and also keep the sparseness, a new loss function is introduced-ε nonsensitive loss function. The new loss function makes SV regression estimation possess robustness and nice sparseness.

ε nonsensitive loss function is as the followings:

$$\begin{cases} L(y, f(\mathbf{x}, \mathbf{w})) = L(|y - f(\mathbf{x}, \mathbf{w})|_\varepsilon) \\ |y - f(\mathbf{x}, \mathbf{w})|_\varepsilon = \begin{cases} 0 & \text{if } |y - f(\mathbf{x}, \mathbf{w})| \le \varepsilon \\ |y - f(\mathbf{x}, \mathbf{w})| - \varepsilon & \text{otherwise} \end{cases} \end{cases} \tag{5.30}$$

The functions are as follows: if the difference between prediction and real value is less than ε, the loss is zero, that is to say, the point in ε range has no effect on loss function. Therefore this loss function assures the solution to possess nice sparseness.

According to ε nonsensitive loss function we know that if the difference between $f(x_i) = (\omega \cdot x_i) + b$ and y_i is not greater than ε, it equals zero when error is negligible. If the difference is greater than ε, the error is: $|f(x_i) - y_i| - \varepsilon$, as is shown in Fig. 5.24.

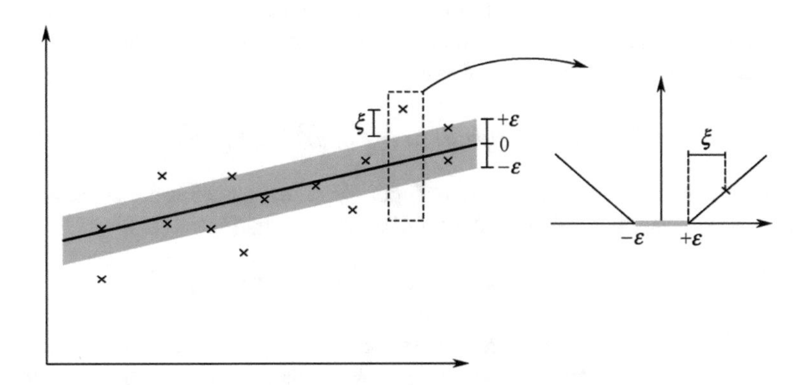

Fig. 5.24 Error diagram of nonsensitive loss function

3. Kernel function

In order to realize SVM in nonlinear situation, we have to utilize nonlinear mapping algorithm of kernel feature space. The basic idea is to map the input to a high-dimensional feature space by using a nonlinear mapping, and then apply linear SVM in the new high-dimensional space.

Nonlinear mapping algorithm of feature space is based on nonlinear mapping: $\Phi : \boldsymbol{R}^n \rightarrow \boldsymbol{F},\ \mathbf{x} \rightarrow \Phi(\mathbf{x})$. The sample data $x_1, \ldots, x_n \in R^n$ is mapped to high-dimensional space F, and the learning algorithm is applied in transformed high-dimension space F. Thus the sample data is converted into

$$[\Phi(x_1), y_1], \ldots, [\Phi(x_n), y_n] \in \boldsymbol{F} \times \boldsymbol{Y} \tag{5.31}$$

According to the functional theory, only if a function satisfies Mercer condition, it is a kernel function of a transformation. The simplified Mercer condition is:

$$\iint K(x, y) g(x) g(y) \mathrm{d}x \mathrm{d}y > 0, \quad f \in L_2(x) \tag{5.32}$$

In the formula: $L_2(x)$ represents L2 (R) spaces (square integrable space); g is the arbitrary function in L2 (R) spaces; $K(x, y)$ is the function in L2 (R) spaces. If the above formula is workable, $K(x, y)$ is the kernel function.

Here are some of the common kernel functions:

(1) Polynomial kernel function: $k(\mathbf{x}, y) = [(\mathbf{x} \cdot y) + 1]^d$
(2) Sigmoid perception kernel function: $K(x, y) = \tanh[k(x \cdot y) + \theta]$
(3) Gaussian kernel function: $K(x, y) = \exp\left(-\frac{(x-y)^2}{2\sigma^2}\right)$

4. Regression algorithm

The basic idea of SVR is to find a nonlinear mapping ϕ from the input space to the output space. Through this nonlinear mapping ϕ, the data is cross mapped to the high-dimensional feature space F, and carry on linear regression of this space in the eigenspace by using the following linear function. The given training sample set is:

$$\boldsymbol{T} = \{(x_1, y_1), (x_2, y_2), \ldots (x_n, y_n)\} \in (\boldsymbol{X} \times \boldsymbol{Y})^n \tag{5.33}$$

In the formula, $x \in R^n$ is the value of input variables, $y \in R^n$ is the corresponding output value, n is the number of the training sample, that is

$$f(x) = (\omega \cdot \varphi(x)) + b \quad \varphi : \boldsymbol{R}^n \rightarrow \boldsymbol{F}, \omega \in \boldsymbol{F} \tag{5.34}$$

In the formula, b is the threshold value. Because φ is fixed, factors which could influence φ are the total empirical risk R_{emp} and ω that makes φ flat in high-dimensional space. Thus functional approximation problem is equivalent to what makes the following functionality to its minimum. That is

$$R_{\text{reg}}\ [f] = R_{\text{emp}} + \frac{1}{2}\|\omega\|^2 = \frac{1}{2}\|\omega\|^2 + \frac{C}{l}\sum_{t=1}^{l} e(f(x_i) - y_i) \tag{5.35}$$

In the formula, l is the number of the samples; $e()$ is the loss function. With the consideration that the linear nonsensitive loss function ε not only has nice sparseness but also can ensure a better generalization capability of the results, consequently nonsensitive loss function ε formula (5.5.35) is applied. Set empirical risk as

$$R^{\varepsilon}_{\text{emp}}\ [f] = \frac{1}{l}\sum_{i=1}^{l} |y - f(\mathbf{x})|_{\varepsilon} \tag{5.36}$$

And expression (5.5.40) is equivalent to the optimization problem solved by the following formula.

$$\min J = \frac{1}{2}\|\omega\|^2 + C\sum_{i=1}^{s}(\xi_i^* + \xi_i) \tag{5.37}$$

$$\text{s.t.} \begin{cases} y_i - (\omega \cdot \Phi(\mathbf{x})) - b \leq \varepsilon + \xi_i^* \\ (\omega, \Phi(\mathbf{x})) + b - y_i \leq \varepsilon + \xi_i \\ \xi_i, \xi_i^* \geq 0 \end{cases} \tag{5.38}$$

Establish the Lagrange equation:

$$\begin{aligned} L(\omega, \xi_i, \xi_i^*) = {} & \frac{1}{2}\|\omega\|^2 + \frac{C}{l}\sum_{i=1}^{l}(\xi_i + \xi_i^*) - \sum_{i=1}^{l} a((\varepsilon + \xi_i^*) + y_i - (\omega \cdot \phi(x_i)) - b) \\ & - \sum a_i^*((\varepsilon + \xi_i^*) - y_i + (\omega \cdot \phi(x_i)) + b) \end{aligned} \tag{5.39}$$

To get the minimum value, partial derivative for parameters $\omega, b, \xi_i, \xi_i^*$ equal zero, that is

$$\begin{cases} \frac{\partial l}{\partial \omega} = \omega - \sum_{i=1}^{l}(a_i - a_i^*) \cdot \varphi(x_i) = 0 \\ \frac{\partial l}{\partial b} = \sum_{i=1}^{l}(a_i - a_i^*) = 0 \\ \frac{\partial l}{\partial \xi_i} = C - a_i - \lambda_i = 0 \\ \frac{\partial l}{\partial_i^*} = C - a_i^* - \lambda_i^* = 0 \end{cases} \tag{5.40}$$

which can be converted into:

$$\min J = \frac{1}{2}\sum_{i,j=1}^{l}(\alpha_i - \alpha_i^*)(\alpha_j - \alpha_j^*)K(\mathbf{x_i}, \mathbf{x_j}) + \sum_{i}^{l}\alpha_i^*(y_i - \varepsilon) - \sum_{i=1}^{l}\alpha_i(y_i + \varepsilon)$$

$$\text{s.t.} \begin{cases} \sum_{i=1}^{l}\alpha_i = \sum_{i=1}^{l}\alpha_i^* \\ 0 \le \alpha_i \le C \\ 0 \le \alpha_i^* \le C \end{cases} \tag{5.41}$$

Theoretically, if the points on boundary satisfy $(0 \le \alpha_i, \alpha_i^* \le 1/\lambda)$, the only definiteness $\delta_k = \varepsilon \operatorname{sign}(\alpha_k - \alpha_k^*)$ of prediction errors could be obtained. Taking into consideration of stability, the average of the boundary point is recommended:

$$b = \text{average}_k\{\delta_k + y_k - \sum_i(\alpha_i - \alpha_i^*)k(\mathbf{x_i}, \mathbf{x_k})\} \tag{5.42}$$

The nonlinear mapping of solving the above convex quadratic programming is expressed as:

$$f(\mathbf{x}) = \sum_{i=1}^{s}(\alpha_i - \alpha_i^*)k(\mathbf{x_i}, \mathbf{x}) + b \tag{5.43}$$

5.4.2 The Weighted Least Square Support Vector Machine Arithmetic (WLS-SVM)

SVM for classification and nonlinear estimation is proposed by Vapnik. As an important achievement in the field of neural networks and nonlinear modeling, SVM is further studied by some other scholars. Many problems, like local minimum and how to select the number of hidden layer, are existed in traditional neural networks methods such as MLP and RBF. SVM is a convex optimization problem and is decided by several extra adjustment parameters. Further, the complexity of the model is also related to convex optimization problem. Particularly, in order to decide SVM model, convex Quadratic Programming (QP) problem might be solved in even space. Optimization expression which is related to QP problem involves indeterminate constraint. When function estimation is carrying on, Vapnik's ε nonsensitive loss function formula is adopted. An interesting feature of SVM method is that it must be sparsely approximated. That is to say, many elements equal zero in QP solution vector. SVM is a kernel-based method, and it allows the kernels to meet the Mercer condition, such as linear kernels, polynomial kernels and RBF kernels.

The major purpose of this section is to solve robustness problem under the existing LS-SVM framework. The present study achieves a better robust estimation with the application of LS-SVM. The process is obtained from relating the weight with error variable on the basis of error variable gained from unweighted LS-SVM.

1. LS-SVM for Function Estimation

In LS-SVM which is proposed by Suykens and Vandewalle, the optimization index uses quadratic term and only has equality constraint instead of quadratic programming. The process of function estimation is as below.

Given N points of training set: $\{\mathbf{x}_k, \boldsymbol{y}_k\}_{k=1}^N$ among them the input is $\mathbf{x}_k \in \boldsymbol{R}^n$, the output is $\boldsymbol{y}_k \in \boldsymbol{R}$. Consider the following optimization problem:

$$\min_{\mathbf{w},\mathbf{e}} J(\mathbf{w},\mathbf{e}) = \frac{1}{2}\mathbf{w}^T\mathbf{w} + \frac{1}{2}\gamma\sum_{k=1}^{N} e_k^2 \tag{5.44}$$

$$\text{s.t.}\begin{cases} y_k = \mathbf{w}^T\Phi(x_k) + b + e_k \\ k = 1,\ldots,N \end{cases} \tag{5.45}$$

Define the Lagrange function:

$$L(\mathbf{w}, b, \mathbf{e}; \boldsymbol{\alpha}) = J(\mathbf{w},\mathbf{e}) - \sum_{k=1}^{N} \alpha_k\{\mathbf{w}^T\Phi(\mathbf{x}_k) + b + e_k - y_k\} \tag{5.46}$$

In the formula, Lagrange multiplier $\alpha_k \in R$. To optimize the formula:

$$\begin{cases} \frac{\partial L}{\partial \mathbf{w}} = 0 \Rightarrow \mathbf{w} = \sum\limits_{k=1}^{N} \alpha_k\Phi(\mathbf{x}_k) \\ \frac{\partial L}{\partial b} = 0 \Rightarrow \sum\limits_{k=1}^{N} \alpha_k = 0 \\ \frac{\partial L}{\partial e_k} = 0 \Rightarrow \alpha_k = \gamma e_k \quad k = 1,\ldots,N \\ \frac{\partial L}{\partial \alpha_k} = 0 \Rightarrow \mathbf{w}^T\Phi(\mathbf{x}_k) + b + e_k - y_k = 0 \quad k = 1,\ldots,N \end{cases} \tag{5.47}$$

Except for $\alpha_k = \gamma e_k$ the above conditions are similar to the standard SVM optimality condition.

With the elimination of w and e, the formula is simplified into:

$$\left[\begin{array}{c|c} 0 & 1_v^T \\ \hline 1_v & \Omega + \frac{1}{\gamma}I \end{array}\right]\left[\begin{array}{c} b \\ \hline \boldsymbol{\alpha} \end{array}\right] = \left[\begin{array}{c} 0 \\ \hline \mathbf{y} \end{array}\right] \tag{5.48}$$

In the formula, $\mathbf{y} = [y_1; \ldots; y_N]$, $1_v = [1; \ldots; 1]$, $\boldsymbol{\alpha} = [\alpha_1; \ldots; \alpha_N]$, $\Omega_{kl} = \Phi(\mathbf{x}_k)^T\Phi(\mathbf{x}_l) = k(\mathbf{x}_k, \mathbf{x}_l)$.

The regression LS-SVM function is

$$f(\mathbf{x}) = \sum_{k=1}^{N} \alpha_k k(\mathbf{x}, \mathbf{x}_k) + b \tag{5.49}$$

In the formula, $\boldsymbol{\alpha}$, b are the solution of formula (5.5.48).

2. WLS-SVM

On the basis of original LS-SVM method, in order to get a better robustness estimation, error variable $e_k = \alpha_k/\gamma$ is measured by defining weight coefficient v_k Given training set: $\{\mathbf{x}_k, y_k\}_{k=1}^{N}$, the input is $\mathbf{x}_k \in R^n$, and the output is $\boldsymbol{y}_k \in \boldsymbol{R}$. It comes down to the following optimization problem:

$$\min_{\mathbf{w}^*, \mathbf{e}^*} J(\mathbf{w}^*, \mathbf{e}^*) = \frac{1}{2}\mathbf{w}^{*T}\mathbf{w}^* + \frac{1}{2}\gamma \sum_{k=1}^{N} v_k e_k^{*2} \tag{5.50}$$

$$\text{s.t.} \begin{cases} y_k = \mathbf{w}^{*T}\Phi(\mathbf{x}_k) + b^* + e_k^* \\ k = 1, \ldots, N \end{cases} \tag{5.51}$$

Define the Lagrange function:

$$L(\mathbf{w}^*, b^*, \mathbf{e}^*; \boldsymbol{\alpha}^*) = J(\mathbf{w}^*, \mathbf{e}^*) - \sum_{k=1}^{N} \alpha_k^* \left\{ \mathbf{w}^{*T}\Phi(\mathbf{x}_k) + b^* + e_k^* - y_k \right\} \tag{5.52}$$

In the formula, Lagrange multiplier $\alpha_k^* \in R$. The unknown parameter of WLS-SVM is expressed as*.

To optimize the above formula:

$$\begin{cases} \frac{\partial L}{\partial \mathbf{w}^*} = 0 \quad \Rightarrow \quad \mathbf{w}^* = \sum_{k=1}^{N} \alpha_k^* \Phi(\mathbf{x}_k) \\ \frac{\partial L}{\partial b^*} = 0 \quad \Rightarrow \quad \sum_{k=1}^{N} \alpha_k^* = 0 \\ \frac{\partial L}{\partial e_k^*} = 0 \quad \Rightarrow \quad \alpha_k^* = \gamma v_k e_k^* \\ \frac{\partial L}{\partial \alpha_k^*} = 0 \quad \Rightarrow \quad \mathbf{w}^{*T}\Phi(\mathbf{x}_k) + b^* + e_k^* - y_k = 0 \end{cases} \tag{5.53}$$

With the elimination of w^* and e^*, the formula is simplified into:

$$\left[\begin{array}{c|c} 0 & 1_v^T \\ \hline 1_v & \Omega + V_\gamma \end{array}\right] \left[\begin{array}{c} b^* \\ \hline \boldsymbol{\alpha}^* \end{array}\right] = \left[\begin{array}{c} 0 \\ \hline \mathbf{y} \end{array}\right] \tag{5.54}$$

In the formula, $\mathbf{y} = [y_1; \ldots; y_N]$, $1_v = [1; \ldots; 1]$, $\boldsymbol{\alpha} = [\alpha_1; \ldots; \alpha_N]$, the diagonal matrix V_γ is :

$$V_\gamma = \mathrm{diag}\left\{\frac{1}{\gamma v_1}, \cdots, \frac{1}{\gamma v_N}\right\} \tag{5.55}$$

The selection of v_k is determined by the error variable $e_k = \alpha_k/\gamma$ of unweighted SVM.

$$v_k = \begin{cases} 1 & \text{if} \quad |e_k/\hat{s}| \le c_1 \\ \frac{c_2 - |e_k/\hat{s}|}{c_2 - c_1} & \text{if} \quad c_1 \le |e_k/\hat{s}| \le c_2 \\ 10^{-4} & \text{otherwise} \end{cases} \tag{5.56}$$

As the usual, $c_1 = 2.5$ and $c_2 = 3$ are reasonable choices. In view of this, no residue value is greater than $2.5\hat{s}$ for Gaussian distribution. Besides, decisions of c_1, c_2 are based on the probability estimation of e_k distribution.

In the formula, $\hat{s}$ is the robustness estimation of standard deviation of WLS-SVM error variable e_k:

$$\hat{s} = \frac{\mathrm{IQR}}{2 \times 0.6745} \tag{5.57}$$

Composition range IQR indicates the difference between 75 and 25%. In making an estimation of $\hat{s}$, the amount of deviation from Gaussian distribution of the estimated error distribution may also be considered. Another robustness estimation of standard deviation is:

$$\hat{s} = 1.483\ \mathrm{MAD}(x_k) \tag{5.58}$$

In the formula, MDA means absolute median deviation. With the application of kernel trick $\Omega_{kl} = \boldsymbol{\Phi}(\mathbf{x}_k)^T \boldsymbol{\Phi}(\mathbf{x}_l) = k(\mathbf{x}_k, \mathbf{x}_l)$ can be obtained, and the WLS-SVM regression function is:

$$f(\mathbf{x}) = \sum_{k=1}^{N} \alpha_k k(\mathbf{x}, \mathbf{x}_k) + b \tag{5.59}$$

5.4.3 WLS-SVM Parametric Prediction Model

1. Determination of the predictor structure

For the given time series $\{x_1, x_2, \ldots, x_N\}$, assuming $x(t)$ is known, with the prediction of $x(t+1)$, the mapping could be constructed $f : \boldsymbol{R}^m \to \boldsymbol{R}$:

$$\hat{x}(t+1) = f(x(t-(m-1)), \ldots, x(t-1), x(t)) \tag{5.60}$$

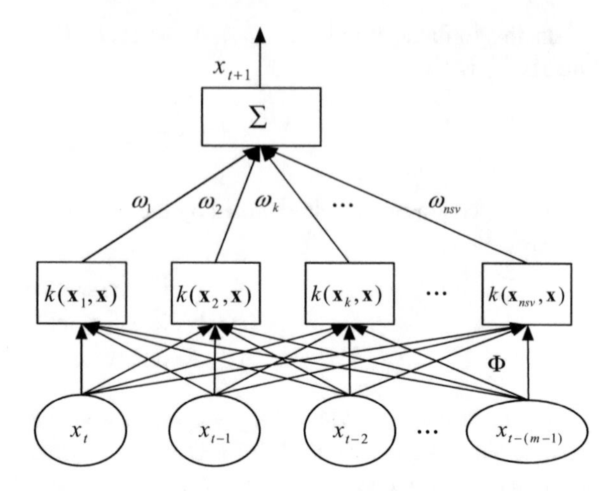

Fig. 5.25 Structure of WLS-SVM predictor

WLS-SVM is used to complete nonlinear mapping approximation. The topology of WLS-SVM predictor structure is shown in Fig. 5.25. In the figure, m is the embedded dimensions, and it is actually the step of the model. Because SVM has the arbitrary approaching nonlinearity mapping capability, and the predictor network structure determined by the aforementioned mapping is generated by automatic optimization algorithm, therefore optimized predictor structure can be obtained by means of suitable model selecting principle to optimize the selection of the number of SVM inputting knots.

2. Determination of the Embedding Dimension

We may set up a model for time series $\{x_1, x_2, \ldots, x_N\}$ and divide it into two parts. The proceeding n_{tr} data is to train the predictor, estimate the parameter and determine the topological structure. The rest of the data is to modify the effectiveness of the model. For the purpose of effective model prediction, phase space should be reconstructed, namely, to transform the one-dimensional data into matrix form so that data relationship is obtained, maximum amount of information is mined out. Set up the mapping relationship $f : \boldsymbol{R}^m \to \boldsymbol{R}$ between input and output $y_t = \{\mathbf{x}_t\}$, the value of embedding dimension m reflects the knowledge quantity implicated in the transformed matrix.

$$\mathbf{x}_t = \{x_{t-1-(m-1)\tau}, \ldots, x_{t-1-\tau}, x_{t-1}\} \tag{5.61}$$

$$\boldsymbol{X} = \begin{bmatrix} x_1 & x_2 & \cdots & x_m \\ x_2 & x_3 & \cdots & x_{m+1} \\ \vdots & \vdots & \ddots & \vdots \\ x_{n_{\text{tr}}-m} & x_{n_{\text{tr}}-m+1} & \cdots & x_{n_{\text{tr}}-1} \end{bmatrix}, \quad \boldsymbol{Y} = \begin{bmatrix} x_{m+1} \\ x_{m+2} \\ \vdots \\ x_{n_{\text{tr}}} \end{bmatrix} \tag{5.62}$$

There is no well defined theoretical basis for the selection of embedding dimension in time series. Prediction error of BIC evaluation model proposed by Akaike is adopted for the present discussion, and m is selected according to magnitude of error.

$$\text{BIC}\ (m) = n_{\text{tr}}\ \ln\ \sigma_a^2 + m\ \ln\ n_{\text{tr}} \tag{5.63}$$

$$\sigma_a^2 = E(a_{n_{\text{tr}}}) = \frac{1}{n_{\text{tr}} - m - 1} \sum_{t=m+1}^{n_{\text{tr}}} \left[x_t - \left(\sum_{i=m+1}^{n_{\text{tr}}} \alpha_i \cdot k(x_i \cdot x_t) + b \right) \right]^2 \tag{5.64}$$

In the formula, n_{tr} is the number of data used for training; m is the embedding dimension needed to be determined. When the value of m increases, the residual σ_a^2 decreases so that BIC reaches its minimum.

3. Prediction model of WLS-SVM

WLS-SVM training could be conducted after determining the topological structure of WLS-SVM predictor and obtaining the learning sample. The regression function obtained is as follow:

$$y_t = \sum_{i=m+1}^{n_{tr}} \alpha_i \cdot k(\mathbf{x}_i \cdot \mathbf{x}_t) + b \tag{5.65}$$

In the formula $t = m + 1, \ldots, n_{\text{tr}}$

$$\mathbf{x}_t = \{x_{t-1-(m-1)}, \ldots, x_{t-2}, x_{t-1}\} \tag{5.66}$$

One-step prediction model is

$$y_{t+1} = \sum_{i=m+1}^{n_{\text{tr}}} \alpha_i \cdot k(\mathbf{x}_i \cdot \mathbf{x}_{t+1}) + b \tag{5.67}$$

In the formula $\mathbf{x}_{t+1} = \{x_{t+1-m}, \ldots, x_{t-1}, x_t\}$.
l-step prediction model is

$$y_{t,l} = \sum_{i=m+1}^{n_{\text{tr}}} \alpha_i \cdot k(\mathbf{x}_i \cdot \mathbf{x}_{t,l}) + b \tag{5.68}$$

In the formula

$$\mathbf{x}_{t,l} = \{x_{t-m}, \ldots x_{t-l}, \hat{x}_{t,1}, \ldots \hat{x}_{t,(l-1)}\},\ t = n_{\text{tr}} + 1, \ldots, N \tag{5.69}$$

In the formula $\hat{x}$ is the value of prediction.

5.4.4 Application Example

1. Evaluation index

The following statistics are applied to estimate the models in order to evaluate the prediction performance of different models.

$$\mathrm{ERROR} = \frac{1}{N - n_{\mathrm{tr}}} \sum_{t=n_{\mathrm{tr}}+1}^{N} \left| \frac{x(t) - y(t)}{x(t)} \right| \tag{5.70}$$

In the formula: $x(t)$ is the prediction value of $y(t)$.

2. Data preprocessing

Extract 500 data from T4, N1 and N2. In order to reduce the modeling error, extraction data must be processed technically. The processed temperature N2, N1 and N2 data series are shown as in Figs. 5.26, 5.27 and 5.28.

Determine the embedding dimensions of SVM and AR models on the basis of BIC and FPE principles. Embedding dimensions' Figure of T4 Temperature, high and low-pressure rotors are shown as in Figs. 5.29, 5.30 and 5.31.

In Fig. 5.29, embedding dimensions of T4 temperature are different, and its BIC and FPE values are changed accordingly. When $n_{\mathrm{AR}} = 38$, FPE value of AR model reaches to its minimum, at present FPE = 0.016164. When $m_{\mathrm{svm}} = 3$, BIC value reaches to its minimum, at present BIC = −1717.1.

In Fig. 5.30, as with T4 temperature, when embedding dimension of low-pressure rotor $n_{\mathrm{AR}} = 37$, FPE value of AR model reaches to its minimum, at present FPE = 0.024728. When embedding dimension $m_{\mathrm{svm}} = 4$, BIC value of SVM model reaches to its minimum, at present BIC = −1723.8.

In Fig. 5.31, when embedding dimension of high-pressure rotor N2 $n_{\mathrm{AR}} = 39$,

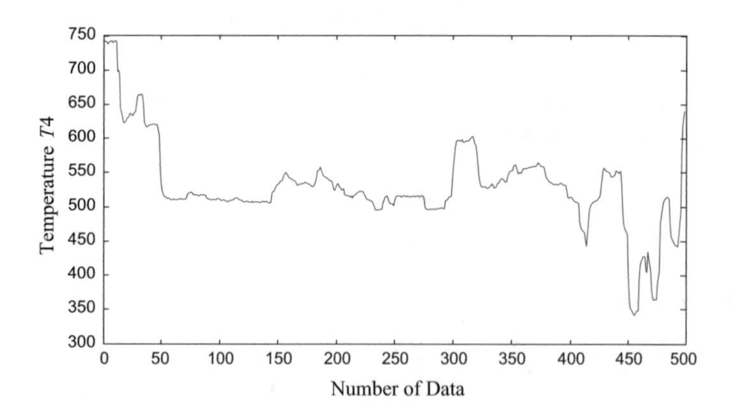

Fig. 5.26 Time series of T4 temperature

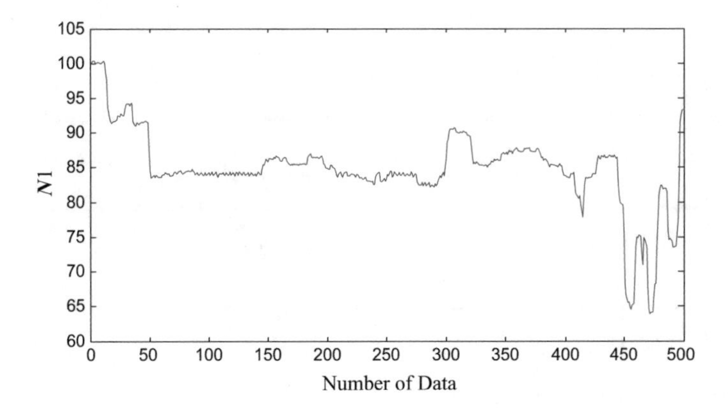

Fig. 5.27 Low-pressure rotor N1 time series

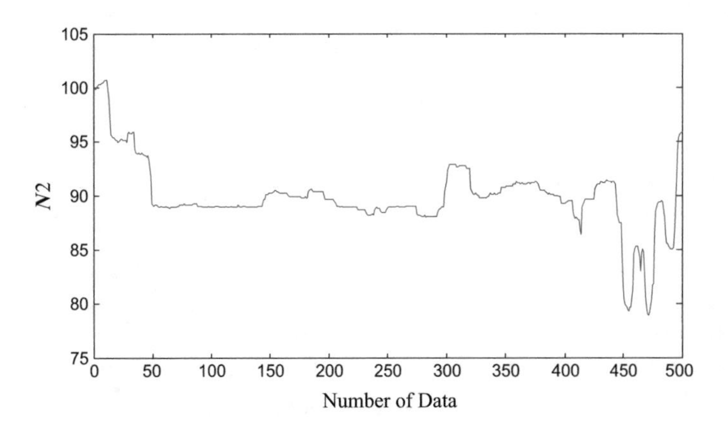

Fig. 5.28 Time series of high-pressure rotor N2

FPE value of AR model reaches to its minimum, at present FPE = 0.021429. When embedding dimension $m_{\mathrm{svm}} = 3$, BIC value of SVM model reaches to its minimum, at present BIC = −1566.4. The Figure also reflects that WLS-SVM regression method has very high modeling accuracy.

3. Experimental results

Divide the data into two groups: take the first 450 data as the training data and the latter 50 data as the testing data. In order to illustrate the effectiveness of WLS-SVM, AR and WLS-SVM prediction model are separately applied to conduct an early 1- to 5-step-prediction. WLS-SVM applies GRBF (Gaussian radial basis function), select $\sigma^2 = 1$, $\varepsilon = 0.005$. Figures 5.32, 5.33, 5.34, 5.35 and 5.36 are the results of T4 temperature conducted by AR and WLS-SVM regression model with

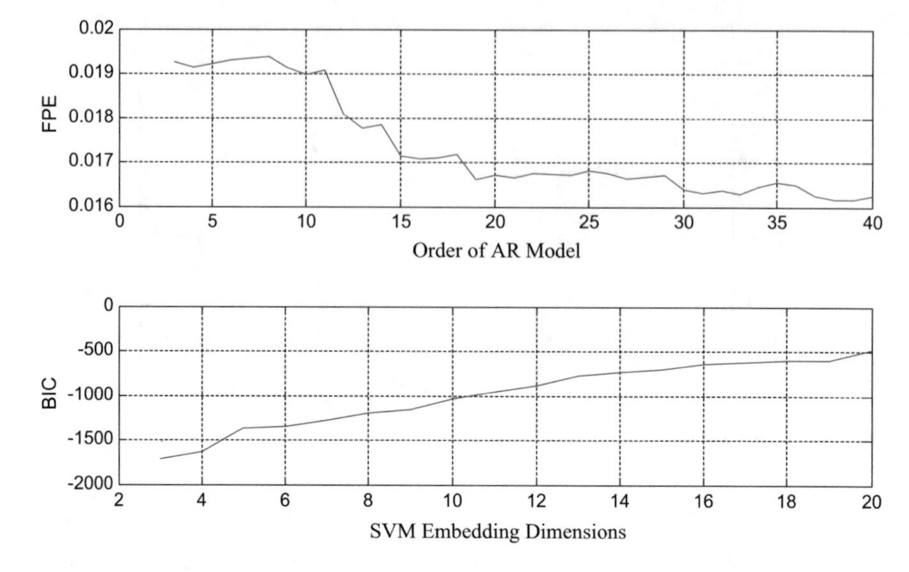

Fig. 5.29 Embedding dimensions of T4 temperature

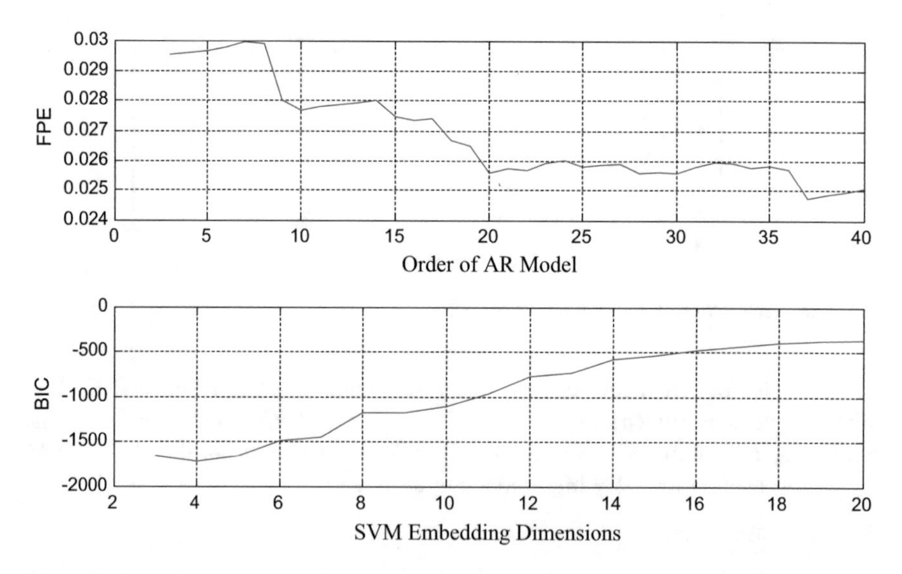

Fig. 5.30 Embedding dimensions of N1

early 1- to 5-step-prediction. The average prediction relative errors of 1 to 5 step are shown as the in Table 5.4.

Table 5.4 shows that AR and WLS-SVM models performed effectively in the short process of T4 Temperature prediction. The 1 step prediction accuracies are 98 and 97.03% with little difference. But in the long process of prediction, WLS-SVM

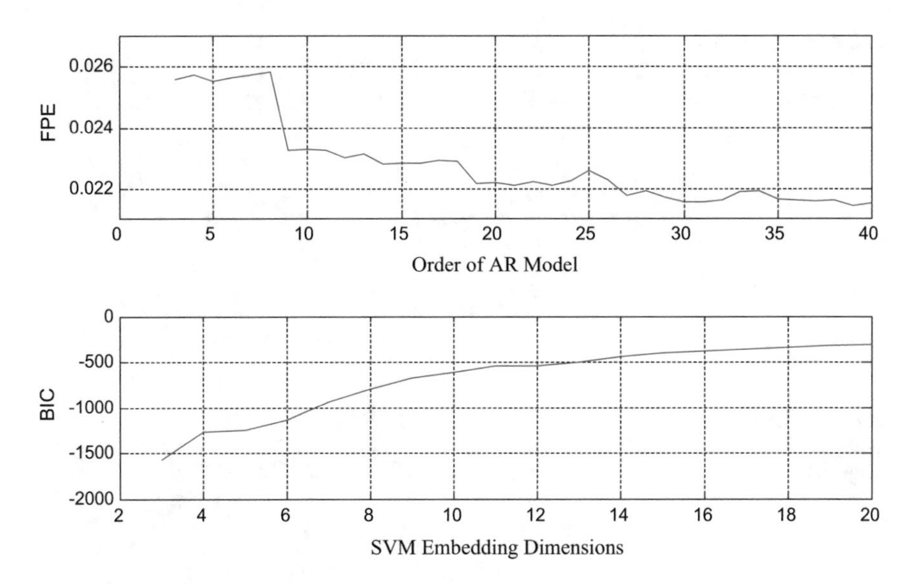

Fig. 5.31 Embedding dimensions of N2

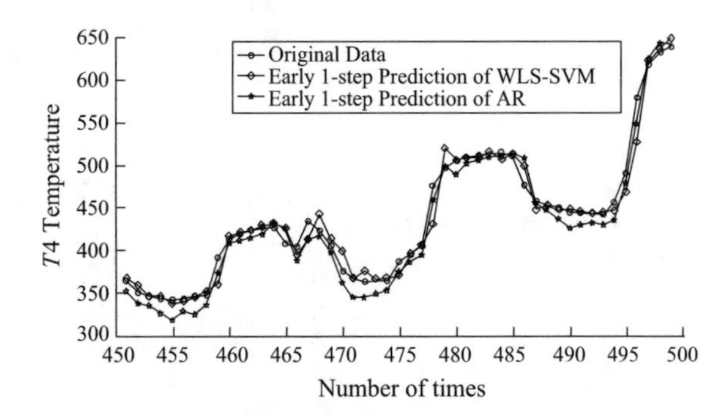

Fig. 5.32 Early 1-step-prediction result of T4 temperature

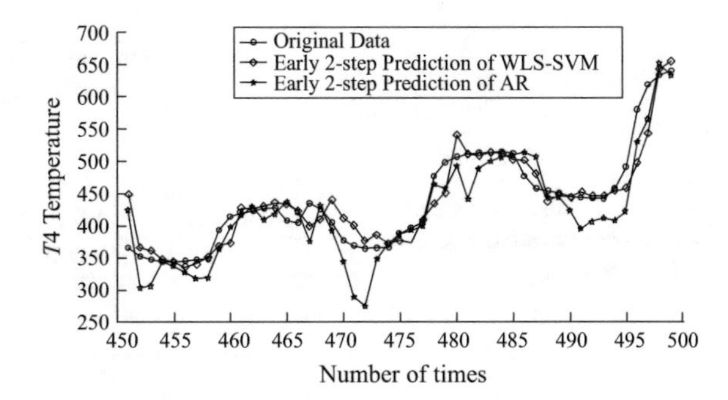

Fig. 5.33 Early 2-step-prediction result of T4 temperature

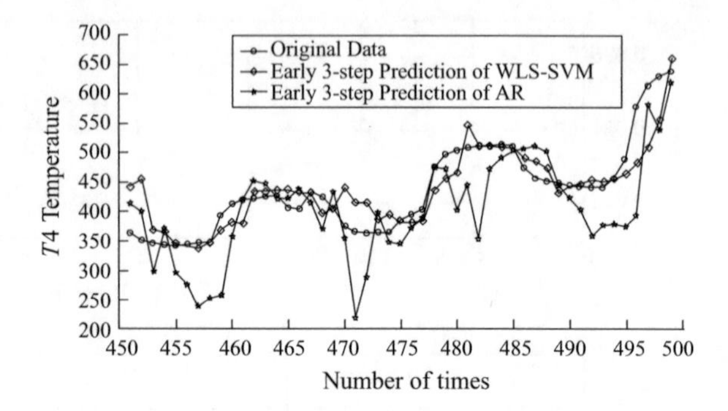

Fig. 5.34 Early 3-step-prediction result of T4 temperature

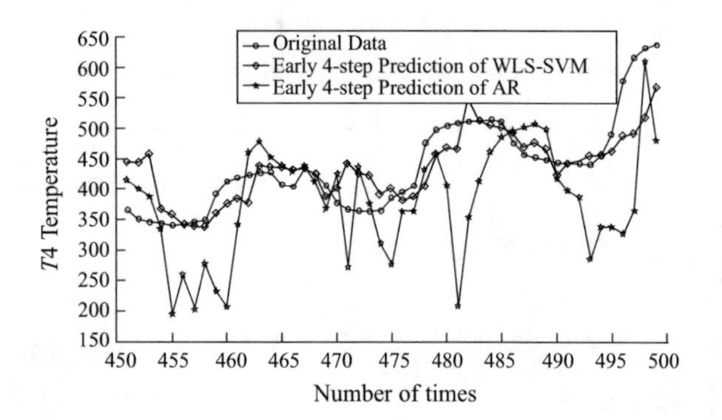

Fig. 5.35 Early 4-step-prediction result of T4 temperature

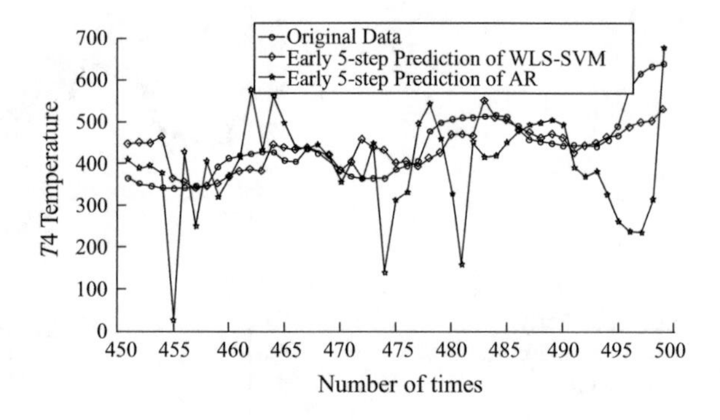

Fig. 5.36 Early 5-step-prediction result of T4 temperature

Table 5.4 Prediction steps and average relative errors of T4 temperature

Prediction steps	WLS-SVM model	AR model
1	0.0200	0.0297
2	0.0428	0.0632
3	0.0601	0.1197
4	0.0767	0.1777
5	0.0911	0.2044

model obviously outperformed AR model. The 5-step-prediction accuracies are 90.89 and 79.56%.

Figures 5.37, 5.38, 5.39, 5.40, and 5.41 are the results of N1 low-pressure rotor conducted by AR and WLS-SVM regression model with an early 1- to 5-step-prediction. The average prediction relative errors of 1 to 5-step are shown as the in Table 5.5.

Table 5.5 shows that both AR and WLS-SVM models achieved high accuracies of 99.27 and 99.47% in the short process of N1 prediction. But in the long process

Fig. 5.37 Early 1-step-prediction result of N1 low-pressure rotor

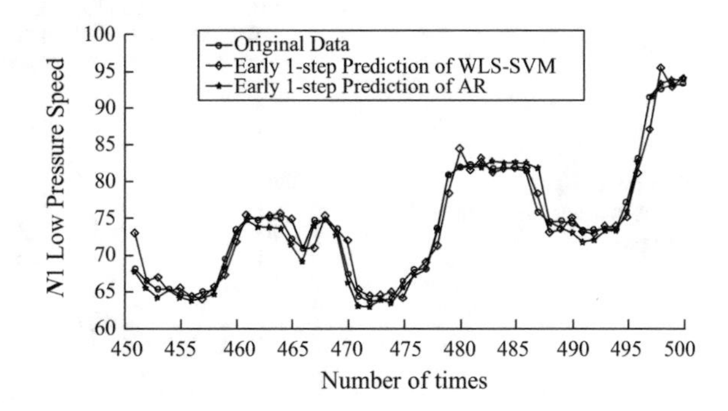

Fig. 5.38 Early 2-step-prediction result of N1 low-pressure rotor

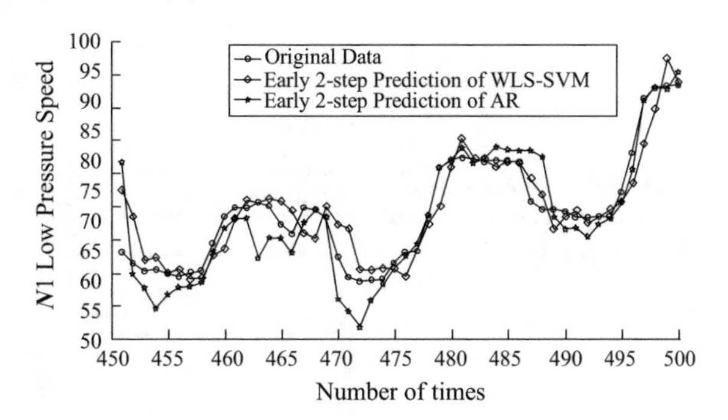

Fig. 5.39 Early 3-step-prediction result of N1 low-pressure rotor

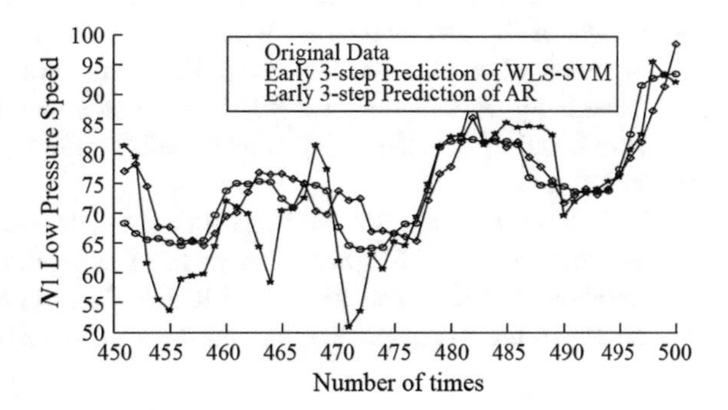

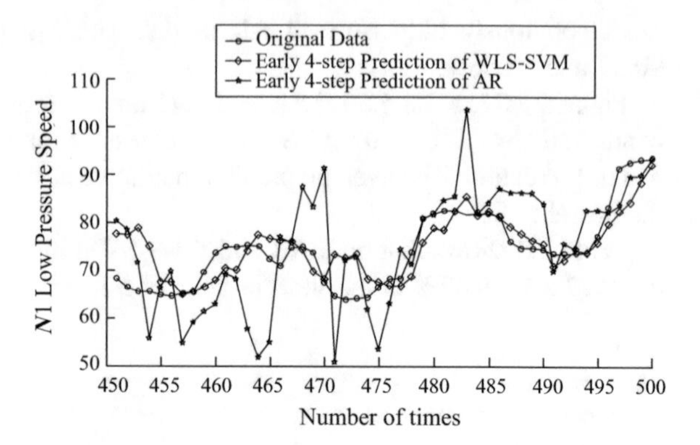

Fig. 5.40 Early 4-step-prediction result of N1 low-pressure rotor

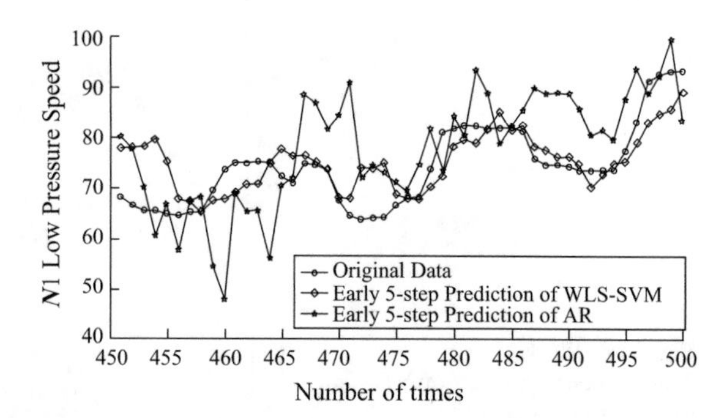

Fig. 5.41 Early 5-step-prediction result of N1 low-pressure rotor

Table 5.5 Prediction steps and average relative errors of N1 low-pressure rotor

Prediction steps	WLS-SVM model	AR model
1	0.0073	0.0053
2	0.0133	0.0318
3	0.0298	0.0797
4	0.0337	0.1175
5	0.0474	0.1474

of prediction, WLS-SVM model obviously outperformed AR model. The prediction accuracies of the two models are 95.26 and 85.26%.

Figures 5.42, 5.43, 5.44, 5.45 and 5.46 are the results of N2 high-pressure speed conducted by AR and WLS-SVM regression model with an early 1- to 5-step-prediction. The average prediction relative errors of 1 to 5-step are shown as the in Table 5.6.

Table 5.6 shows that AR and WLS-SVM models performed with little difference in the short process of N2 prediction. In the 1-step-prediction the accuracies of the two models are 98.29 and 98.83%. AR and WLS-SVM models performed quite differently in the 5-step-prediction, and the accuracies are 94.13 and 88.25%.

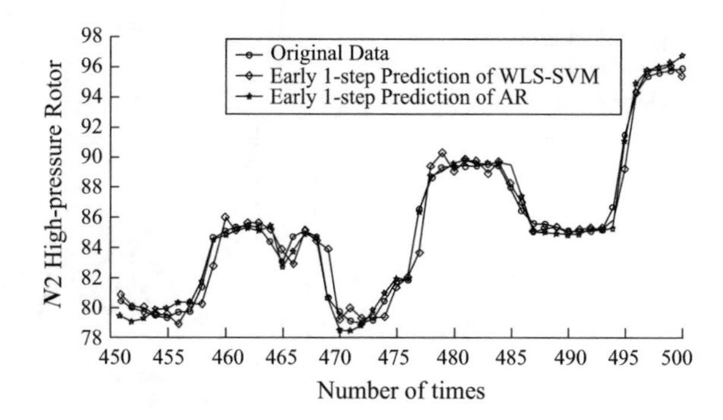

Fig. 5.42 Early 1-step-prediction result of N2 high-pressure rotor

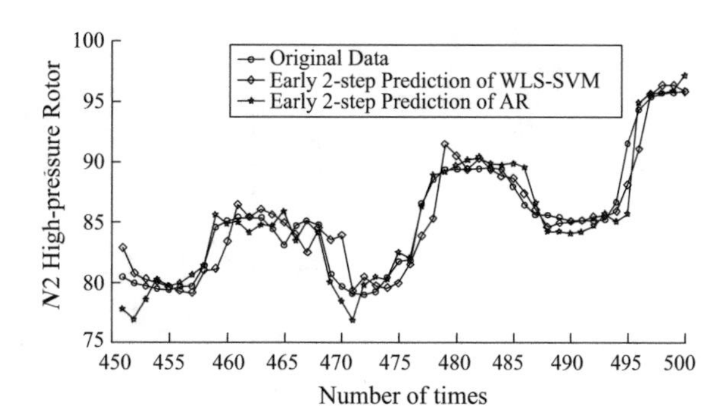

Fig. 5.43 Early 2-step-prediction result of N2 high-pressure rotor

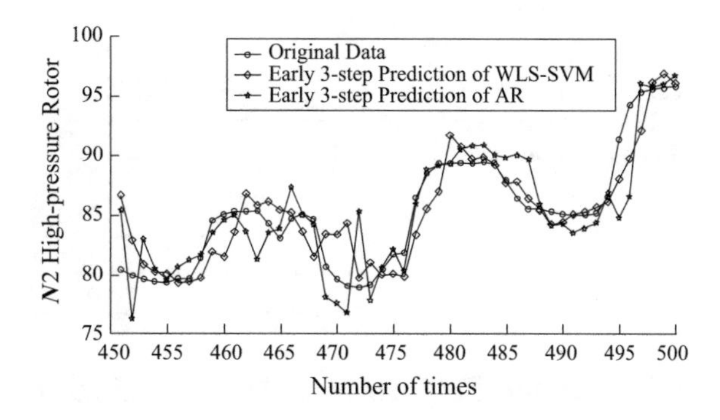

Fig. 5.44 Early 3-step-prediction result of N2 high-pressure rotor

Comprehensively compare the result of the single and multistep prediction, WLS-SVM model obviously outperformed AR model. This conclusion has important significance in the monitor of the flight data. To set up the modeling of flight data in the application of SVM model on the basis of the known flight data series, effective prediction might be conducted five sampling time intervals in advance. If the flight data errors of predication fall into the allowable range, early measures might be conducted to achieve the monitoring goals.

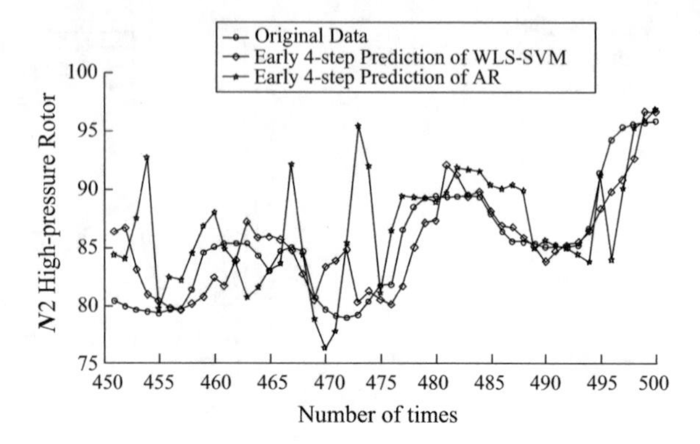

Fig. 5.45 Early 4-step-prediction result of N2 high-pressure rotor

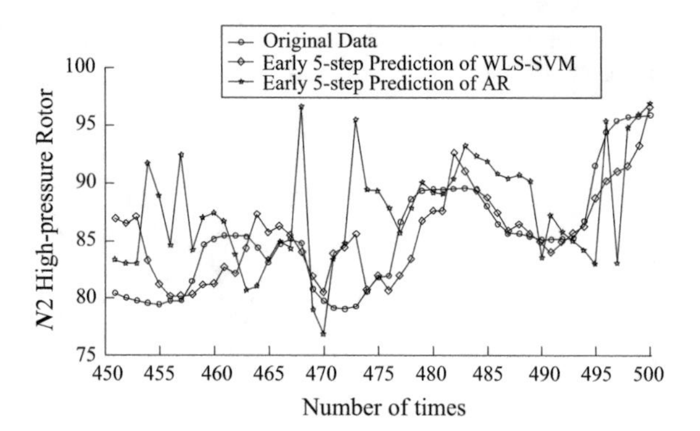

Fig. 5.46 Early 5-prediction result of N2 high-pressure rotor

Table 5.6 Prediction steps and average relative errors of N2 high-pressure rotor

Prediction steps	WLS-SVM model	AR model
1	0.0171	0.0117
2	0.0325	0.0389
3	0.0433	0.0665
4	0.0512	0.1059
5	0.0587	0.1175

5.5 Engine Performance Data Prediction on the Basis of Chaotic Sequence

As the key component of the aircraft, engine performance monitor and fault diagnosis are the crucial part of the maintenance control study, including analysis of engine data tendency, engine gas path, vibration and lubricant. Presently the two effective engine monitoring methods are feature-parameter-based engine performance monitor and model-based engine performance monitor. According to the chaotic feature of flight data, a new prediction method, with the combination of chaotic theory and SVM and other statistical learning theories, is developed. Section 5.5 is to introduce this new method which is applied to predict the typical features of aero -engines.

5.5.1 Chaos and Chaotic Sequence

1. Features of Chaos

(1) Inner randomness. Steady states of chaos are not the general three kinds of states which define the motions-stillness, periodic motion and quasi-periodic motion. Instead they are complicated motions, which are always confined to limited ranges, and never repeat.

(2) Long-time unpredictability. Because the original condition is restricted to a limited accuracy, and the slight difference might greatly influence the later time evolution, dynamic features beyond a certain time in the future cannot be predicted for a long time.

(3) Sensitive dependence for the initial data. As time goes on, any close initial condition show independent time evolution—the sensitive dependence for the initial value.

(4) Universality. When the systems tend to be chaotic, the displayed features are universal. And the systems are not changed as the specific systems and systems' kinematic equations are different.

(5) Fractal. Fractal is the geometric property of the point set in n-dimensional space. The structure of the point set, which is unlimitedly accurate, has the characteristics of self-similarity and overall-similarity. And if the point set has non-integer dimensions which are less than n-dimensional in the space, it is fractal.

(6) Ergodicity. Also called as hybridity, because chaos is complicated motions which always confines to finite region and never repeats. Thus, as time goes chaotic motion trails never linger at a certain condition but traverses at every point in the regional space.

(7) Boundedness. Its motion trajectory is always confined to a certain region which is called chaotic domain of attraction. Generally, chaotic system is stable.

(8) Fractal-dimensionality. The motion state of chaotic system is multivalent and multilayer. The valent and layer are more and more specific to show infinite hierarchical self-similar structure.

(9) Statistic. For chaotic system, positive Lyapunov exponent means the trajectory is unstable in every range, and the adjacent tracks are separated according to the components. Because of the boundedness of attractors, the tracks can only be recurrently folded but never intersect, and the special structure of chaotic attractors is formed.

2. Concept and Nature of Chaotic Sequence.

With the development of nonlinear and chaotic theory, research of chaos is concentrated into application area, especially the application of communication, because the genus-randomness of chaotic system is well suited to noise camouflage debugging. Above all, as chaotic system is sensitively depend on initial phase, tremendous number of signals which are nonrelevant, genus-random but determinate, easily generated and regenerated can be provided.

There are two types of chaotic systems: time sequential system illustrated by differential equation and time discrete system illustrated by state difference equation. The previous one is often used in synchronized chaotic method of secret communication; the latter one is the chaotic sequence which is going to be focused on.

A discrete time dynamic system is defined as

$$x_{k+1} = f(x_k), \quad 0 < x_k < 1, \; k = 0, 1, 2, \ldots$$

In the formula, x_k is the state; $f(x_k)$ maps the present state x_k to the next state x_{k+1}. Iterate from the initial value x_0 to get the sequence:

$$\{x_k : \; k = 0, 1, 2, \ldots\}$$

It is a track of the discrete time dynamic system, also called a chaotic sequence. Natures of chaotic sequence are as follow:

(1) Its production and receiving can be controlled by the user.
(2) It is only produced in the nonlinear (nonlinear mapping) system
(3) Only a certain parameter space corresponds to chaotic state.
(4) Different system initial values lead to completely different chaotic sequence.
(5) Cross correlation of different sequence produced from different chaotic system is almost zero.

5.5.2 *Prediction Model*

Chaotic analysis-based SVM prediction model is to process the phase points in the chaotic phase space. Adopt phase space to reconstruct history flight data, and train the reconstructed phase points by using SVM. The model obtained after training is used to predict engine performance.

(1) First, reconstruct the phase space of flight data, and get the Lyapunov exponent.
(2) Adopt SVM regression flight data to predict the model.

$$y = f(x) = \sum_{i=1}^{k} (a_i - a_i^*) k(x_i - x_j) + b \tag{5.71}$$

In the formula, x_i is the ith sample of k sample; $k(x_i - x_j)$ is the kernel function, the reconstructed phase points are the input samples to support SVM, the specific steps are as follow:

Step 1: From the aforementioned (1), establish multidimensional phase space of chaotic time sequence to construct learning sample and prediction value.
Step 2: Establish target function by using training samples.
Step 3: After the solution with the adoption of modified sequential minimization, the obtained parameters and a_i^* and a_i are substituted in formula (5.5.71).
Step 4: Predict the flight data of a or several certain future moments by using the samples.

(3) Evaluation standard of the model. In order to study the nature of chaotic time sequential prediction model, the following error statistics are used to evaluate the model:

$$\text{ERROR} = \frac{1}{N - n_{\text{tr}}} \sum_{t=n_{\text{tr}}+1}^{N} \left| \frac{x(t) - y(t)}{x(t)} \right| \tag{5.72}$$

In the formula: $y(t)$ is the prediction value of $x(t)$.

5.5.3 *Application of the Model*

1. Parameter selection of the prediction model.

Parameter selection of the prediction model mainly involves the determination of maximum prediction step, monitoring parameter warning threshold and SVM parameter.

(1) Determination of the maximum prediction step.

The maximum prediction step is $T_{\max} = 1/\lambda + 1$, substitute the solved result of the maximum Lyapunov exponent (See details in Sect. 5, Chap. 2) to elicit: maximum prediction step of T_4 temperature $T_{t4\max} = 7$; maximum prediction step of N1 low-pressure rotor $T_{\text{N1}\max} = 6$; maximum prediction step of N2 high-pressure speed $T_{\text{N2}\max} = 14$.

(2) Determination of the monitor parameter warning threshold.

For the two primary methods, the first is given according to the experts' practical maintenance experiences, another is determined by adopting mathematical statistics. The basic thought of the latter method is to assume active engine's characteristic parameters under the same condition submit to Gaussian distribution.

(3) Determination of SVM parameter

Chaotic sequential based SVM is used to conduct performance prediction. In order to test the performance of this method, WLS-SVM is also used in the prediction. Besides, error comparisons between these two methods are given. Equality constraints in WLS-SVM is different with the inequality constraints in traditional SVM, because LS linear system is used as the loss function but not the quadratic programming method which is used in traditional SVM. Thus the complexity of the modeling is effectively reduced.

The application of SVM parameter model is mainly to determine the suitable penalty factor C, loss function nonsensitive factor ε and radial basis function σ^2. The method which is used in this book is to set the value of σ^2 and C as 10 and 100, the corresponding value of ε is gradually increased from 0.001, the MSE of training set and test set is stable and not influenced by the value ε. Select ε as 0.001.

During the selection of specific selection, when C and ε are separately fixed to 100 and 0.001, the training standard MSE increases as σ^2 increases. Meanwhile as σ^2 increases, the standard MSE the test set first decreases and later increases. It shows that if the value of σ^2 is too small, "over learning" of the training set might be caused. And if the value of σ^2 is too big "under learning" of the training set might be caused. Dereferencing of σ^2 should be among 1–100. When C and ε are separately fixed to 10 and 0.001, the training standard MSE decreases monotonously as C increases. Meanwhile the standard MSE the test set decreases gradually. When the value of C increases from 10 to 200, the standard MSE the test set almost maintain a constant. When the value of C increases to 200, the standard MSE the test set begins to increase. Dereferencing of C should be among 10–200.

2. Single-step prediction on the basis of chaotic sequence.

The aforementioned theory is used to predict the engine's T4 temperature, the High and low-pressure rotor. Specifics are as follows.

(1) Embedded dimension of the phase space is determined according to the GP algorithm. Embedded dimension of T4 temperature is 6, embedded dimension of low-pressure rotor is 4, embedded dimension of high-pressure rotor is 5.
(2) Phase space reconstruction is adopted for data, and set up multidimensional phase space. Take the previous 450 data as the learning samples, and take the latter 50 data as the test samples.
(3) SVM adopts RBF kernel function, and select the suitable penalty factor, loss function nonsensitive factor and radial basis function. The values of them are: $C = 100$, $\sigma^2 = 10$, $\varepsilon = 0.001$.

(4) Predict the next-time data according to the single-step prediction model and test data. Each test is based on the test data. Only to predict the next data point. Single-step prediction results of T4 temperature, high and low-pressure rotor are separately shown in Figs. 5.47, 5.48 and 5.49. Figures show that single-step prediction has very high accuracy. Chaotic theory achieves very good effect, and realizes effective data prediction.

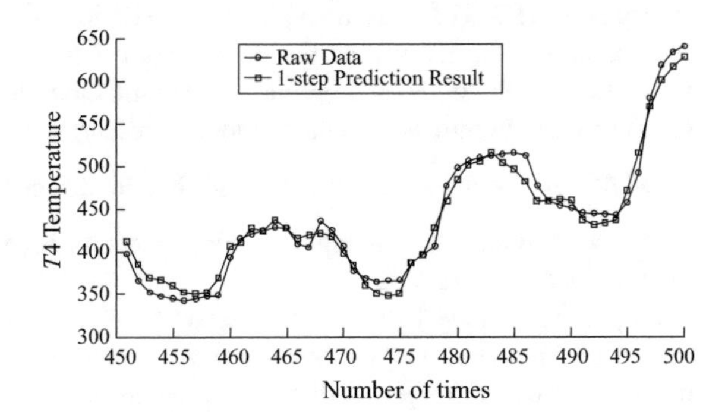

Fig. 5.47 Single-step chaotic prediction result of T4 temperature

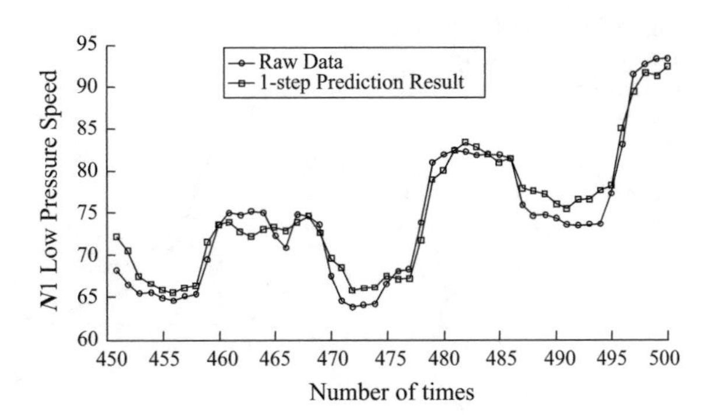

Fig. 5.48 Single-step chaotic prediction result of N1 low-pressure rotor

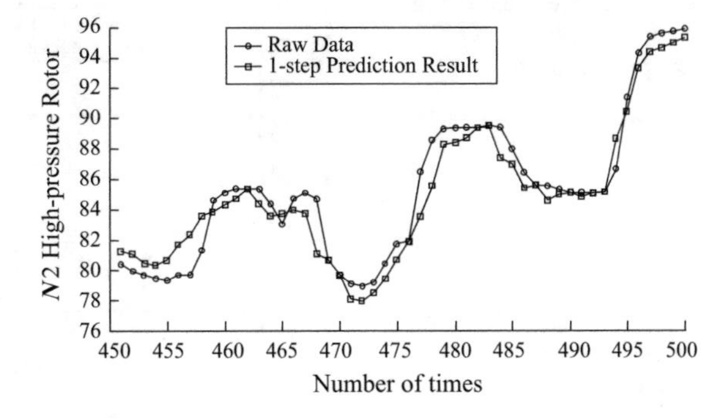

Fig. 5.49 Single-step chaotic prediction result of N2 high-pressure rotor

Table 5.7 Single-step prediction error

	SVM based on chaotic theory	WLS-SVM
T4 temperature	0.0017	0.0200
N1 low-pressure speed	0.0091	0.0073
N2 high-pressure speed	0.0045	0.0171

Table 5.7 compares the errors of SVM single-step prediction based on chaotic theory and WLS-SVM based single-step prediction. The result shows that with the consideration of relativity of the data chaotic based SVM method perform nice generalization in single-step prediction. The method also provides theoretical base for the future aircraft performance monitor research.

3. Multistep prediction on the basis of chaotic sequence.

In this section, 1- to 15-step-predictions are applied, and the results are shown in Figs. 5.50, 5.51, and 5.52.

Figure 5.50 shows that the real effective prediction steps agree basically with the maximum prediction steps estimated from Lyapunov exponent. Compare with the real data, errors of the proceeding nine prediction points of T4 prediction result are

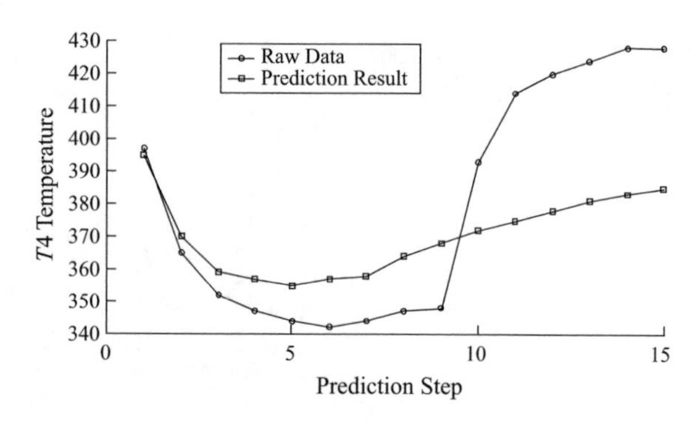

Fig. 5.50 Multistep chaotic prediction result of T4 temperature

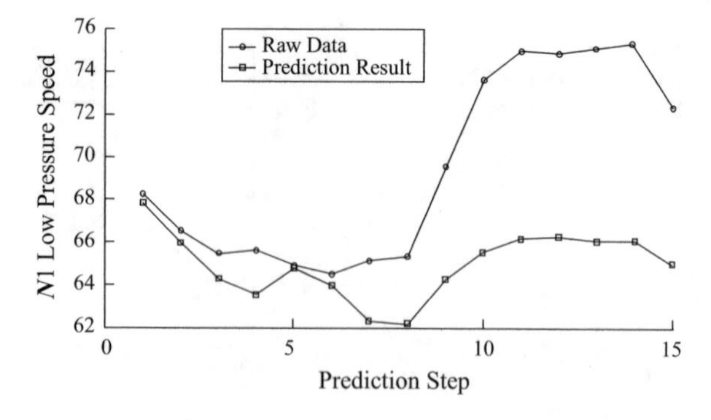

Fig. 5.51 Multistep chaotic prediction result of N1 low-pressure rotor

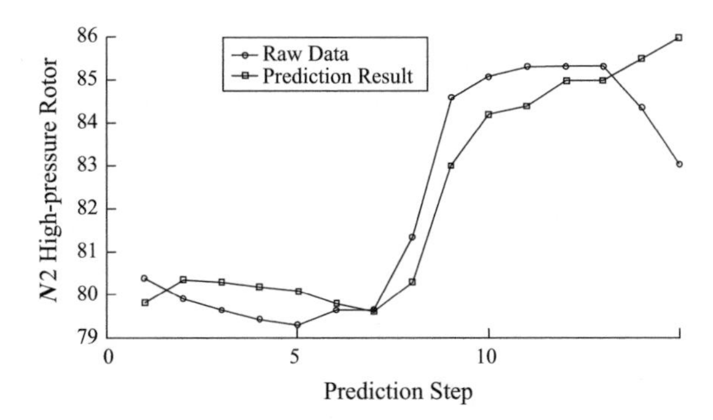

Fig. 5.52 Multistep chaotic prediction result of N2 high-pressure speed

small. But the values of errors begin to increase greatly from the 10th data, and at the moment the prediction model is invalid.

Figure 5.51 shows that with the comparison of the real data, errors of the proceeding eight prediction points of N1 prediction result are small. The effective prediction steps agree basically with the maximum prediction steps estimated from Lyapunov exponent.

Figure 5.52 shows that with the comparison of the real data, errors of the proceeding 14 prediction points of N2 prediction result are small. The effective prediction steps agree basically with the maximum prediction steps estimated from Lyapunov exponent. From the 15th data, the values of errors begin to increase greatly.

The above results show that chaotic system possesses initial sensitivity. With the accumulation of errors, even though single-step and several-step prediction results of the prediction model approximate to the real value, errors could make output value of the system completely deviate from the real value in the future recursive multistep prediction. Therefore, multistep prediction of chaotic system could approximate to the real value within a certain time range. When the time range is exceeded, the system could not achieve a long term accurate prediction.

Chapter 6
Design and Implementation of Flight Data Mining System

In this chapter, a flight-data-based prototype system of data mining is designed and proposed, which is an instance of practical application of the research findings of the present study in this book. The prototype system takes advantage of the current flight data processing systems and their engineering applications, making effective use of applicable data mining systems, DBMiner for instance. It is capable of performing data query and statistics, as well as "mining out" patterns of time series. Data patterns to be mined out include those of operation conditions of aircraft parts and units and aircraft maneuverings. In addition to being capable of condition monitoring and predicating, and in particular, it works very well in engine health and trend prediction, and therefore promises a wide range of application.

6.1 Data Mining System

Data mining, also known as **knowledge discovery from data** (KDD), is an advanced data process procedure aiming at searching for credible, innovative, effective, and understandable patterns in data from a great deal of raw material. By incorporating methods such as statistics, fuzzy mathematics, neural networks, expert systems, and machine learning, data mining extracts abstract knowledge from a huge amount of data so as to reveal the essential rules and internal relations of the objective world that are hiding in the data. The ultimate purpose of data mining is automatic knowledge acquisition.

Currently, the research of data mining focuses on preprocessing for data mining, process of data mining, and post-processing of data mining (including expression and interpretation of mining results), which are also points of departure of the research. The research covers the following topics: frequent patterns, association rules, classification rules, clustering patterns, anomaly patterns, and periodic regularity as well. To be more specific, association rules are used to identify

J. Zhang and P. Zhang, *Time Series Analysis Methods and Applications for Flight Data*, DOI 10.1007/978-3-662-53430-4_6

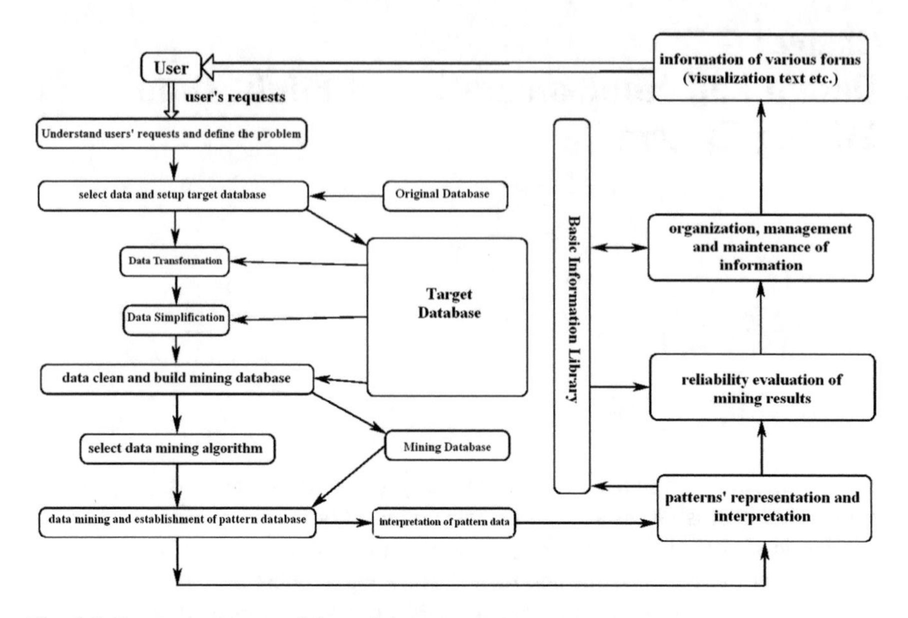

Fig. 6.1 Prototype structure of data mining system

association or correlation among a great number of items of data sets; classification rules are used to extract data models that describe important data or to predict data trend; clustering patterns are used to classify the object data into types or clusters when data of unknown types are classified; abnormality analysis aims at identifying object data that conform with patterns of general behavior, or models; and periodic trend analysis is done to identify in the object data the trends and rules that change over time. The structure of the prototype system is illustrated in Fig. 6.1.

As illustrated in Fig. 6.2, the working of a typical module of data mining system normally goes like this: ① storing accumulated data all together in the management system of a relational database using applicable data mining systems or modules;

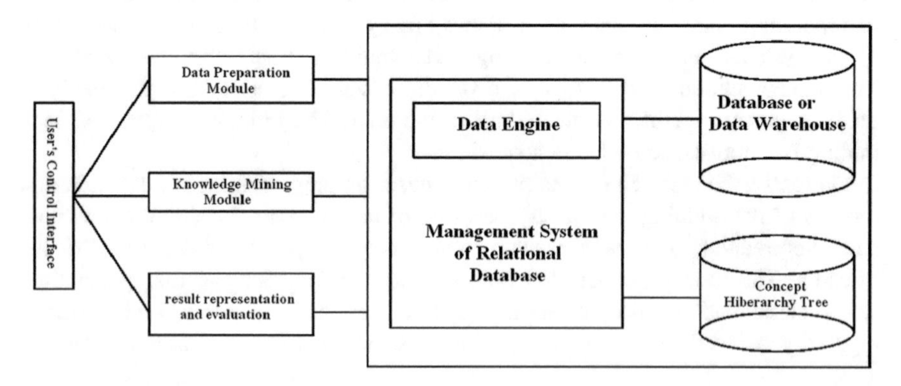

Fig. 6.2 Typical modules data mining system

② extracting a large amount of relational data by the engine of database or data warehouse, which is capable of effective data management, query, and analysis; and ③ performing the function of data management and analysis by the data engine.

The advantages of the above-mentioned data mining system mainly lie in its reduction of functional redundancy, improvement of system consistency, and seamless space owning to its capability of managing massive spatial data without the necessity of space partitioning. These advantages enable the system to perform effective spatial analysis and data mining in a way that the logical continuity of spatial data is guaranteed. Currently, in spite of the achievements in the application of data mining, many problems remain unsolved, among which the following two are critically important to the practical application of data mining.

(1) The study of complicated relationships in multi-attribute and massive data. Currently, a majority of data mining systems are no more than two-dimensional data tables that support manual extraction only even though they are based on relational database. Therefore, for massive data mining from large databases, it is essential to study the complicated relationships between higher dimensions and attributes.
(2) The establishment of easily understandable information representation and interpretation mechanism. In practical application of data mining system, the central problem is the understandability of the extracted information to the user. However, the current research on the method of knowledge representation has only touched the shallow depth of the issue. For example, the current methods of data visualization can do no more than a simple graphic description of the results of data mining, failing to reveal the essence of the method.

This chapter aims at solving the two problems. Through flight data warehouse modeling, establishment of data mining models, as well as the use of effective relational search and visualization technology, effective mining of flight data and overall monitoring of engine condition will be achieved.

6.2 Flight Data Warehouse Modeling

6.2.1 The Special Characteristics of Flight Data

Flight data are special mainly in the following two ways:

(1) Flight data are measured and obtained in a chronological order and at equal time intervals; and a flight database is normally made up of flight data series values or events that change over time. In addition to storing time series data, a flight database can also perform such operations to the stored data as selecting,

projecting, connecting, as well as inquiring for various purposes. The analysis of times series data may result in reliability-based prediction of data ranges. As to the prediction itself, it follows a three-step procedure: determining the target data curve, extracting the model order of the corresponding time series, and establishing regression models.

(2) The state points of flight data are characterized with nonrepeatability, that is to say, system state represented by the points is very unlikely to repeat itself. Flight data are recorded frame by frame; the data recorded are parameters of various types, including position, motion, operation, primary system condition, and alarm. Except for time of flight and times of engagement, most of flight data are nonadditive, namely, a simple addition of data does not make any mathematic sense. The mining of single-frame parameter targets at interrelations between parameters; for example, synchronization rate between parameters during aircraft operation, lag time, and threshold of difference values. These interrelations can be used for judgment of control sensitivity and operation stability. The study of parameter distribution helps in evaluation of system health and warning of a possible failure.

6.2.2 *Mining Goals of Flight Data*

Tasks of data mining mainly cover concept description, association, classification, prediction, clustering, time series analysis, and so on. As far as data analysis is concerned, these tasks can be classified into two categories: descriptive and predictive data mining. The former, which is often referred to as concept description, describes the data in a brief way and make generalizations of them; the latter performs data analysis for the purpose of establishing one or a group of models, and in an attempt to predict the pattern of behavior of new data sets.

Concept description is not simply an enumeration of data, but rather it generates characterization and comparison of data. Data characterization provides the given data sets with a clear summary, while data comparison supplies comparative descriptions of two or more data sets. The association rules are used to "dig out" the relationship between items in the given data sets in a process that involves two steps: ① finding out all the frequent item sets, the frequency of which should at least be the same in count as the predefined minimum support degree; ② on the basis of the frequent item sets, generating strong association rules which must satisfy minimum support degree and confidence. Data classification also follows a two-step process: first, building a model that is able to describe predetermined sets of data classes or concepts, a step also known as supervised learning; second, evaluating the prediction accuracy of the models, and using them for classification. Different from classification, clustering classifies unknown classes of data, and is therefore an unsupervised learning or observational learning.

In the present study, flight data mining puts its emphasis on multidimensional data models in an attempt to perform quick and effective query to the data in the models with respect to concept hierarchy, data attributes, and index relations, etc. This kind of flight data mining enables effective mining of such data patterns as periodic patterns, patterns of aircraft working state, and patterns of aircraft maneuvering; and, mining of these patterns makes possible similarity analysis. Moreover, flight data mining as such also enables monitoring and prediction of parameters of basic characteristics, and of operation conditions as indicated by aircraft operation models.

6.2.3 Flight Data Warehouse Modeling

As shown in Fig. 6.3, since flight data processing is characterized with multigranularity, the following steps should be taken in the building of flight data warehouse:

(1) Establishment of multidimensional data models. Normally, star connection is adopted for the establishment of flight data warehouse so as to process and optimize data for decision support system. Through preconnection, a selective redundancy of data can be established.
(2) Concept hierarchy. Concept hierarchy injects into itself background knowledge, and dual granularity is usually adopted in the actual process of injection. The corresponding hierarchical structure is jointly provided by system users and experts or knowledge workers in the domain of flight data. Concept hierarchy can be divided into several subtypes, including mode hierarchy, set-grouping hierarchy, export operation hierarchy, and rule-based hierarchy.
(3) Selection and application of measurement methods. In online analytical processing (OLAP), and among other statistic measurements, both terms of central tendency measurement and terms of data dispersion are used; the former includes terms like *mean*, *median*, *mode*, and *midrage* while the latter includes *quartile*, *outlier*, and *variance.*

Flight data warehouse data and operational data are different in the following ways. ① The time limit of data warehouse is much longer than that of operational data systems, the former expected to be 5–10 years, while the later only 60–90 days. ② Operational databases contain data of the "current value" the accuracy of which is valid only at time of visit; data of the same current value will be updated, while warehouse data are only a series of complex snapshots generated at a certain moment. ③ The key code structure of operational data may or may not include time elements such as year, month, and day, while that of warehouse data always contains time elements. A multidimensional data model flight data is shown below in Fig. 6.4.

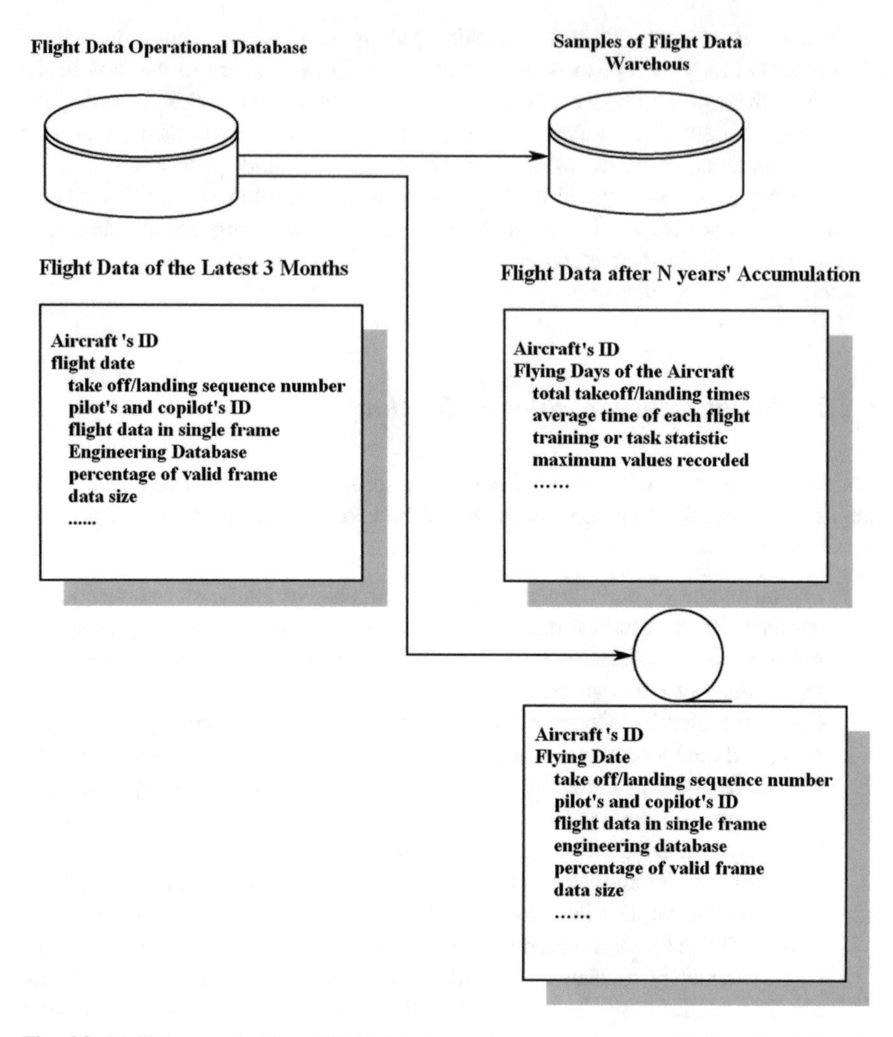

Fig. 6.3 Multiple granularities of flight data processing

6.3 Design and Development of Prototype System

6.3.1 General Design

The analysis and design of prototype system follows the principle of modularization in software engineering; the basic thinking goes like this:

(1) Quick reading of distributed flight data: original data, quickly processed data, and detailedly processed data with reports, all should find their quick entry into the data warehouse.

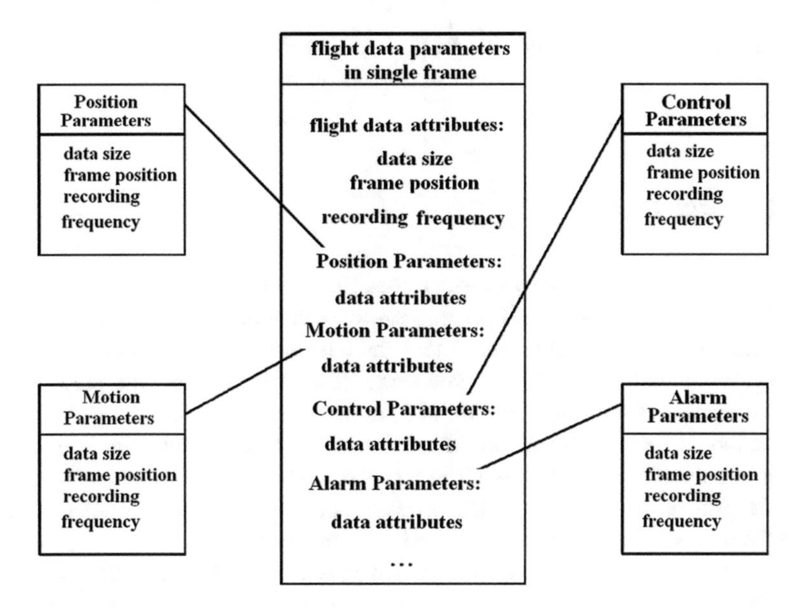

Fig. 6.4 A multidimensional data model of flight data

(2) Building data warehouse capable of quick adjustment: the flight data warehouse should be built according to the multidimensional flight data model and different concept granularity, involving adjustment of co-text information, establishment of index, and refreshing of the external and unstructured data in module memory.

(3) Performing effective mining of patterns of time series: the data mining involves information processing, on line analysis, and pattern mining, and it should also support similarity search and basic statistical analysis. OLAP provides basic operations in the data warehouse, data slicing for example, and supports multidimensional data analysis incorporating graphical representation of time series. In the mining of aircraft maneuvering patterns, clustering analysis based on temporal continuity is enabled.

(4) Reusing reported results and mode of mining: in order for the model to be supplied, the association rules "dug out" from the reports must first be refiltered by the metarules, and then restrictions are attached to the results of refiltering. The model obtained is used to dig out effectively the categories of aircraft maneuvering and patterns of working status of aircraft systems. By means of reusing the mode of mining, patterns of the same category can be dug out. The reusing is realized by making use of class inherited technology in the object-oriented technology.

(5) Monitoring and predicting of aircraft working status: one of the important applications of flight data mining is monitoring and predicting of aircraft operation condition, including state monitoring, short-time prediction, intelligent analysis of prediction results, decision-making suggestions, graphical

output of prediction and trends, as well as the representation of decision-making suggestion in the form of a decision tree.

6.3.2 Data Flow

The data flow of the prototype system goes through the following steps: first, by means of a data connection and preprocessing module data sets and their attributes are interactively selected from the data warehouse for visualized classification and grading, which is followed by preprocessing of the data; second, the preprocessed data are displayed in various views; and third, at the same time the decision tree training module is introduced into the process of decision tree training. The training results include the decision-making tree before and after cutting, as well as some classification rules. The knowledge of Bayesian network can be learned directly from the data sets; or it can be considered as prior knowledge according to the rules of the decision-making tree, and be learned in relation to user's understanding of the data. In his way, graphic representation of Bayesian network can be obtained.

Since the intermediate results of data mining support real-time visualization, users may further analyze the inherent features of the training data sets, and consider modifying the training data sets or the two types of data mining parameters, prior knowledge, processing results to improve the overall effect learning and training. When the users find that the current clustering rules satisfy their needs, they may also select from the data warehouse data sets for testing according to the clustering rules. The test results can be displayed in the map along with the training data so that the two data sets can be compared with respect to their spatial distribution and classification. Following the testing, interactive clustering can be done to the rest unknown data. There can be a switch at any time between the three processes, as well as between the different views so that there will be a circular and progressive process of knowledge discovery. Figure 6.5 is an illustration of the relations between the data flows in different parts of the system.

6.3.3 Working Process

The establishment of the working process begins with input/output data structure of the target system; and for this purpose, a data-structure-oriented method of software development is adopted. The working process itself begins with building of data cube by extracting relational data from flight data warehouse in the way as required by the user. Then, a preliminary concept hierarchy is done on the basis of the data cube. And then, knowledge mining is done by employing association rules,

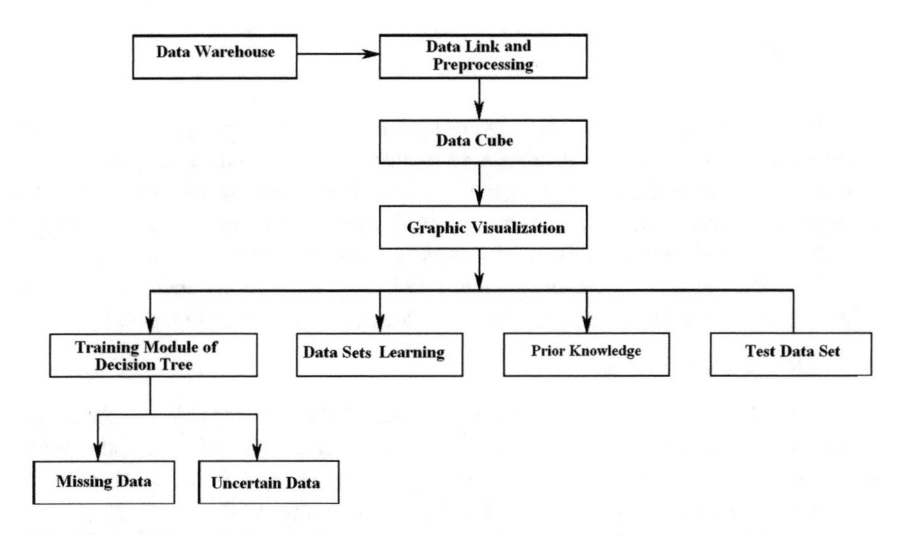

Fig. 6.5 Data flow relations in the system

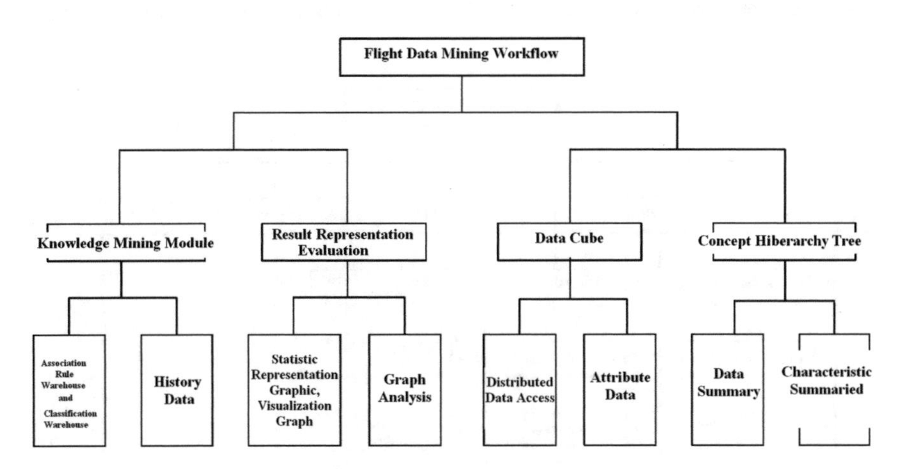

Fig. 6.6 System data mining workflow

classification algorithm, and clustering algorithm; the discovered knowledge and patterns are transferred to the display module. For the sake of convenience and easy understanding, representation of the working process depends much on graphic descriptions statistic summary, for example, tables, histograms, quantile plots, quantile–quantile plots, scatter plots, local regression curves, dot plots, etc. The process of mining is shown in Fig. 6.6.

6.3.4 Main Functions

The mainstream user interface (UI) is adopted for the prototype system. The UI is designed in such a way that it provides functions of form input, menu, and direct interaction, with the emphasis on visualized output of the results of data mining. In order for the goals of the design to be achieved, as shown in Fig. 6.7, the prototype system is divided into the following functional parts: flight data warehouse, data query, pattern mining, monitoring and predicting, as well as reuse of pattern mining, etc. The main system interface and menu are shown in Fig. 6.8.

1. Flight Data Warehouse

The flight data warehouse built for the present study is capable of automatic unloading, transmitting, and importing the flight data. Meanwhile, the data warehouse is accompanied by a corresponding flight event store. A flight event refers to any of such events as takeoff, landing, flight endurance, and aircraft operations close to performance and power limits. And the events are indicated by the parameters obtained through fast processing of the flight data. Flight events together with their parameters prove to be very helpful to quick and effective monitoring of

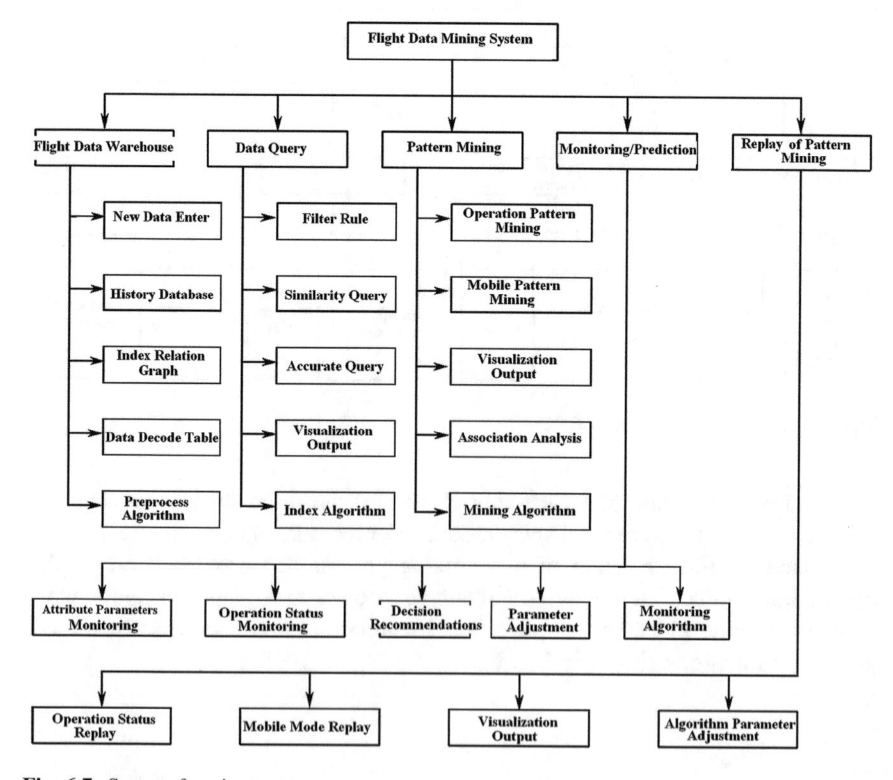

Fig. 6.7 System functions

Data entry (V)
History Data View (W)
Index Relation(X)
Data Decode (Y)
Preprocess Algorithm(Z)
Exit (E)

Filter Rules (V)
Similarity Query (W)
Accurate Query (X)
Visualization Output (Y)
Index Algorithm (Z)

Operation Mode Mining (V)
Mobile Pattern Mining (W)
Visualization Output (X)
Corelation Analysis(Y)
Mining Algorithm(Z)

Attribute Data Prediction (V)
Operation Prediction (W)
Decision Recommendation(V)
Data Adjustment (Y)
Prediction Algorithm(Z)

Operation Mode Reuse (W)
Mobile Mode Reuse (V)
Visualization Output (X)
Algorithm Data Adjustment (Z)

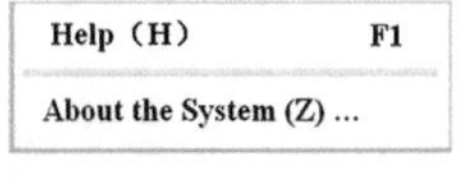

Fig. 6.8 System main interfaces and menu

aircraft health. At the same time, other operations to the database may also be performed, creating tables and data writing for example, so that the data can be shared among other modules. Moreover, the data warehouse is capable of reading flat files the format of which requires that time series data be separated by a carriage return, and that multiple symbols be used to separate neighboring elements of the data series. As to the construction of index, by means of processing and transforming data noise, extraction of feature vectors and data transformation are done so that an index can be constructed for the entire database of time series. The index data are stored in the files for future query. Some of the functions are illustrated in Figs. 6.9 and 6.10.

2. Similarity Search of Flight Data Series

Flight data series similarity search may be divided into whole series search and subseries search. In the former, the whole database is the search for all the data series that satisfy the query condition; while in the latter, the series that satisfy the query condition may occupy any position in any data series in the database, and they may be of arbitrary length in fuzzy query. In terms of the characteristics of query, similarity search can also be divided into search based on numerical distance between

选择信息[Q] 模 拟 量[P] 开 关 量[S] 发动机常数[M] 帮 助[H] 退 出[X]

符号	全名称	类型	最小值	最大值	格式	单位	频率	插值点
ATAKA	迎角	模拟量	-7.0	40.0	4.1f	o	8	11
BHAl	左发alfa1	模拟量	-2.0	110.0	5.1f	code	4	11
BHAr	右发alfa1	模拟量	-2.0	110.0	5.1f	code	4	11
BZl	左发斜板	模拟量	0.0	100.0	5.1f	%	2	5
BZr	右发斜板	模拟量	0.0	100.0	5.1f	%	2	5
Bck	侧滑角	模拟量	-10.0	10.0	5.1f	o	4	14
CTl	左平尾	模拟量	-20.0	15.0	5.1f	o	4	8
CTr	右平尾	模拟量	-20.0	15.0	5.1f	o	4	8
EKPAH	故障代码	辅助参数	0.0	255.0	3.0f	code	2	0
FLl	左襟副翼	模拟量	-35.0	20.0	5.1f	o	4	11
FLr	右襟副翼	模拟量	-20.0	35.0	5.1f	o	4	11
GT	剩油量	模拟量	0.0	10000.0	6.0f	Kg	2	13
HAPICl	左发alfa2	模拟量	-2.0	120.0	5.1f	code	4	11
HAPICr	右发alfa2	模拟量	-2.0	120.0	5.1f	code	4	11
HOCl	左前襟	模拟量	-2.0	30.0	4.1f	o	4	5
HOCr	右前襟	模拟量	-2.0	30.0	4.1f	o	4	5
HOUR	小时	辅助参数	0.0	23.0	2.0f	H	1	0
Hb	气压高度	模拟量	-250.0	25000.0	6.0f	m	3	24
Hcbc	大气机高度	数字量	-500.0	30000.0	6.0f	m	3	1
KPEH	倾斜角	数字量	-180.0	180.0	6.1f	o	12	1
KYPC	航向角	数字量	0.0	360.0	5.1f	o	3	1
L1901	参数组1	辅助参数	0.0	255.0	3d	code	2	0
L1902	参数组2	辅助参数	0.0	255.0	3d	code	1	0
L1903	参数组3	辅助参数	0.0	255.0	3d	code	2	0
L1904	参数组4	辅助参数	0.0	255.0	3d	code	2	0
L1905	参数组5	辅助参数	0.0	255.0	3d	code	2	0
L1906	参数组6	辅助参数	0.0	255.0	3d	code	2	0
MINH	分钟	辅助参数	0.0	59.0	3.0f	MIN	1	0
N1l	左发N1	频率量	5.0	110.0	5.1f	%	2	0
N1r	右发N1	频率量	5.0	110.0	5.1f	%	2	0

地	址
10	42
74	106
138	170
202	234

代码	物理量
5	35.00
28	35.00
51	35.00
75	31.00
98	26.00
122	21.00
145	16.00
170	11.00
195	6.00
226	0.00
248	-4.00

Fig. 6.9 Flight data description

Fig. 6.10 Flight data database table

data series, and search based on mode of change of data series. It is the latter that has been adopted and verified in the present study. In similarity search based on mode of change, since equal-granularity segmentation can be made to data series of different lengths and different value ranges, stability and consistency subsegment are ensured

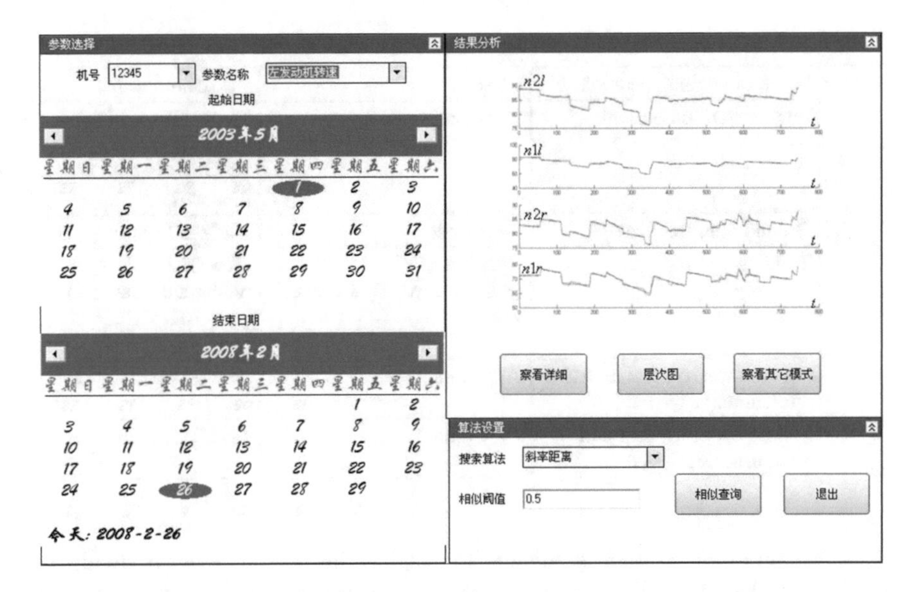

Fig. 6.11 Similarity search of flight data series

and the range of follow-on query is narrowed. This results in not only a higher efficiency of search, but also better validity and operability of the search results. The functional effect of the search method is illustrated in Fig. 6.11.

3. Mining of Patterns of Aircraft Condition

In terms of function, the data mining of the prototype system can be divided into mining of patterns of aircraft operation state, mining of patterns of aircraft maneuvering, and relational analysis. For different states of operation of the type of aircraft under investigation, there are different classifications of modes of operation. Take the engine system for example, there is long list of patterns of the engine operation, including ground start (engine test included), partial augmentation, rating, full throttle, afterburning, takeoff, cruise, and landing, as well as the other patterns at different altitudes or different Mach numbers. All these modes of operations can also be combined to produce more complicated modes. For a specific mode of operation, typical state parameter can be used to monitor the working of the engine. Table 6.1 is a list of the association rules of five modes of engine operation in *T*4 state.

Mode of maneuvering mainly refers to the types of maneuvers performed by a fighter in flight, which can be divided into two major categories: training mode and combating mode. Aircraft maneuvers are often seen as arts of flying, but they must be performed within performance and power limits of the aircraft. In order to win and survive in air combat operations, actually performed aircraft maneuvering is a skillful and complicated combination of basic maneuvers. As indicated by its application to the mining of patterns of aircraft maneuvering, with accuracy over

Table 6.1 Association rules of five modes of engine operation in *T*4

No.	Mode of engine operation	*T*4 state	Association rules
1	Engine start, on ground	*T*4 > limit	$T4 \geq 605 \times \mathrm{dt} \geq 3$ (cold start) $T4 \geq 635 \times \mathrm{dt} \geq 3$ (hot start)
2	Ground idle, stabilized	*T*4 > limit	$T4 \geq 520 \times \mathrm{dt} \geq 5$
3	Training	Duration of *T*4 (> limit) > predetermined value	$T4 > 690 \times \mathrm{dt} \geq 5$ (without afterburner) $T4 > 705 \times \mathrm{dt} \geq 5$ (with afterburner)
4	Maxi. power, combating and afterburner stabilized	*T*4 < limit	$T4 < \mathrm{TD} \times \mathrm{dt} \geq 3$
5	Maxi. power, training and afterburner stabilized	*T*4 < limit	$T4 < \mathrm{TGI} \times \mathrm{dt} \geq 3$

90%, the prototype system is suitable for engineering applications. It should be pointed out that the mining results of aircraft maneuvering patterns can be directly used for training quality evaluation and testing flying skills. The functional effects of this module of the system are illustrated in Figs. 6.12 and 6.13.

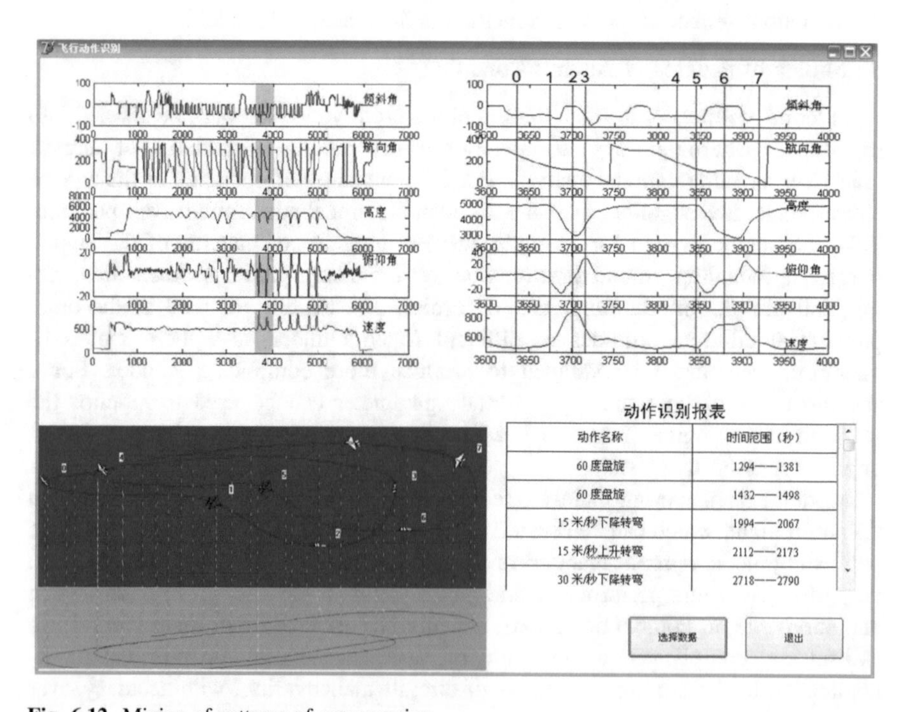

Fig. 6.12 Mining of patterns of maneuvering

飞行事件

事件类型	事件名称	开始时间	结束时间	输出参数	判据
参数统计	飞行中最大表速	0:08:00	0:38:53	1402	!(TMP4)*(TMP1)*(VI>TMP105) 持续1秒
参数统计	飞行中最小表速	0:08:00	0:08:00	258.1081	!(TMP4)*(TMP1)*(VI<TMP205)*(VI>0) 持续1秒
参数统计	飞行中最大载荷因数	0:08:00	0:39:56	2.9497	!(TMP4)*(TMP1)*(NY>TMP117) 持续1秒
参数统计	飞行中最小载荷因数	0:08:00	0:10:18	0.1914	!(TMP4)*(TMP1)*(NY<TMP217) 持续1秒
参数统计	飞行中最大高度	0:08:08	0:21:49	4661	!(TMP4)*(TMP1)*(H>TMP104) 持续1秒
参数统计	加力和最大状态...	0:45:03	0:45:59	2	(TMP6)*(TMP7) 持续1秒
参数统计	空中飞行时间	0:45:03	0:45:59	2043	(TMP6) 持续2秒
参数统计	发动机总工作时间	0:45:03	0:45:59	2391	(TMP6) 持续2秒
飞行状态	起飞	0:07:59	0:08:00	0	!(TMP1)*(N1]90)*(N2]90)*(VI]250) 持续2秒
飞行状态	着陆	0:42:02	0:42:03	0	!(TMP4)*(TMP11)*(VI[310)*(TMP5) 持续2秒

参数统计

参数名称	单位	最大值	最大值时间	最小值	最小值时间
航向角	o	360	347	0	2366
俯仰角	o	53.438	329	-90	357
倾斜角	o	180	0	-179	2754
气压高度	M	4661	1309	-500	357
表速	Km/h	1402	2333	0	0
排气温度	℃	750.221	465	300	0
大气温度	℃	151	0	6.877	2034
调节锥位置	mm	78.221	2705	60	0
力臂臂值	mm	100	357	0	279
交流36伏电压	V	210.82	0	0	270
汇流条电压	V	121.84	2758	7.89	357
流逝时间	s	2765	2759	2	0
侧滑角	o	0.005	328	-0.273	343
平尾偏角	o	24.696	2538	-10.577	442
侧向过载	g	0.392	2529	-1.32	357
轴向过载	g	0.675	468	-0.416	2536
法向过载	g	2.95	2396	0.191	618
油门手柄位移	o	64.062	446	0	212
高压转子转速	%	114.857	600	0	0
低压转子转速	%	100	443	0	0
耗油累计	l	1116.7	2693	0	305
攻角	o	10	286	0	0
方向舵偏角	o	25	207	-25	2691

Fig. 6.13 Flight events report table

4. Integrated Monitoring of Engines

Integrated monitoring discussed in this section is mainly concerned with the monitoring and prediction of engine operation of a type of aircraft. Meanwhile, the discussion also covers another type of aircraft with respect to its control system, engine system, engine accessory system, and hydraulic system. The monitoring and prediction follow the following steps:

(1) Extract from flight data warehouse the data corresponding to the operation modes at different phases so that data needed for modeling can be generated. For the model to be built, call into use the learning module in the model of support vector machine regression. Put in storage the newly built engine state model under the corresponding directory.
(2) Extract the flight data to be monitored, and, by making use of the premade parameter model, call into use the parameter prediction module in the model of support vector regression machine so that the estimated value of corresponding parameters can be generated and that state monitoring and trend analysis can be done.

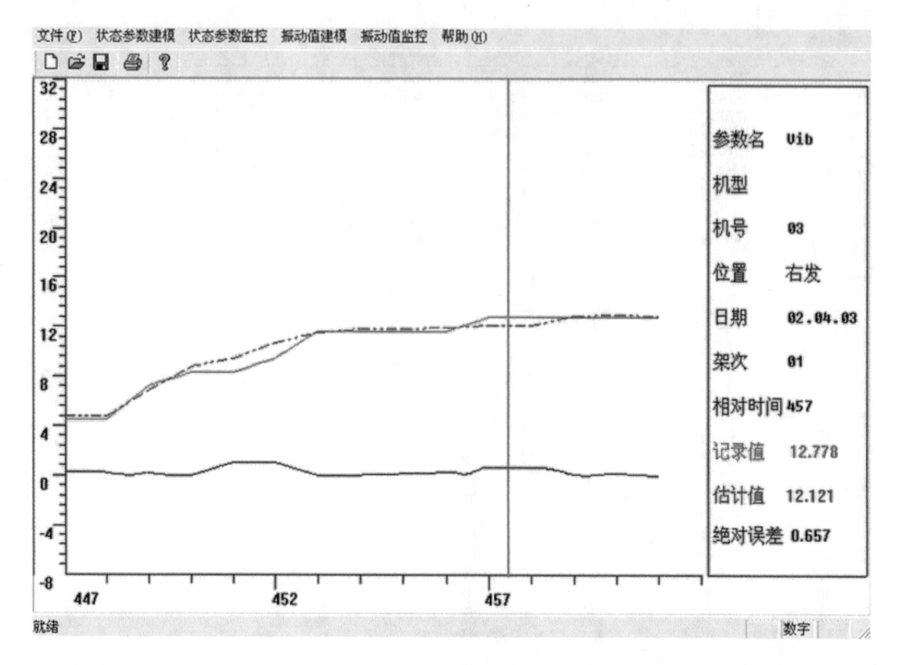

Fig. 6.14 Engine vibration curve in dynamic acceleration phase in multiple sorties

(3) Display graphically such information of the corresponding parameters as estimated value, actual value, absolute error, and time and sortie so that the data files of the corresponding parameters can be generated. Monitor such engine state parameters as *N*2, *N*l, *T*4, and Vib (Engine Vibration Signal) in order to determine whether the engine is in normal condition. Detection of any abnormal parameter will trigger alarming.

It should be noted that in order to ensure accuracy of the model, a new model for engine operations must be built in the case of engine performance adjustment or replacement. Figure 6.14 is a recording of monitoring results of the right engine with respect to its Vib values in continuous sorties.

The effect of parameters trend monitoring has been verified in successfully shooting a hidden trouble in the lubricant system that was likely to cause a flight accident.

6.4 Summary

This chapter is a brief introduction of the general framework of the data mining system. A flight-data-oriented prototype system of data mining is designed and proposed. The system is capable of data query and statistics, and mining of patterns such as aircraft working status and aircraft maneuvering so that aircraft operation state can be monitored and predicted. Engineering applications of the system proves that it works very well in predicting operation and health condition of aircraft engines. In view of its good performance in building of data warehouse, mining of aircraft maneuvering patterns, and monitoring of aircraft conditions, this prototype system can be put into practical use.

References

1. Ni S (2005) Research about high performance aircraft condition monitoring. Northwestern Polytechnical University, Xi-an
2. Li X, Dujun, Zhang P. Aircraft voice system and its application technology. National Defence Industry Press, Beijing
3. Zhang J (2003) Research about aircraft condition and monitoring based on flight data. The Air Force Engineering University, Xi-an
4. Ding Y, Yang X, Chen G (2011) Radian-distance based time series similarity measurement. J Electron Inf Technol 33(1):122–128
5. Xie C (2004) Research about flight data intelligent processing and intelligent processing system development. The Air Force Engineering University, Xi-an
6. Huang Y (2002) Data processing of aircraft condition monitoring system and technology of application. Nanjing University of Aeronautics and Astronautics, Nanjing
7. Liang J (2006) Statistical pattern recognition methods of flight data. The Air Force Engineering University, Xi-an
8. Zhang J (2009) Research about flight data model of multivariate time series processing method. Northwestern Polytechnical University, Xi-an
9. Zhang P (2009) Research about flight data comprehensive development method and application. The Air Force Engineering University, Xi-an
10. Xie C, Ni S, Zhang Z (2005) A new method for preprocessing in absent flight parameter. Comput Simul 22(4):27–31
11. Li G, Li X, Li J (2007) Synthetical filter arithmetic for flight data. Electron Opt Control 14(2):69–72
12. Miao Y, Sun Z (2010) Research on pre-processing method of satellite borne GPS measurement data. Acta Aeronautica Et Astronautica Sinica 31(3):602–607
13. Pan LQ, Li JZ, Luo J (2010) A temporal and spatial correlation based missing values imputation algorithm in wireless sensor networks. Chin J Comput 33(1):1–11
14. Zhu J (2003) The developing tend of aeroengine control. Aircr Des 1:58–61
15. Guo Z, Zhou A (2002) Research on data quality and data cleaning: a survey. J Softw 13(11):2076–2082
16. Galhardas H, Florescu D, Shasha D et al (2001) Declarative data cleaning: language, model and algorithm. In: Proceedings of the 27th international conference on very large data bases. Morgan Kaufmann, Roma, pp 371–380
17. Zhang H, Li X, Kong T (2007) Application of BP neural network in predicting absent data. Comput Eng Des 28(14):3457–3459
18. Hu S, Zhang Z (2007) Fault prediction for nonlinear time series based on neural network. Acta Automatica Sinica 33(7):744–748
19. Li Y, Ni S, Zhang Z (2002) An intelligent approach to flight data processing based on a fuzzy Kohonen neural network. Syst Eng Electron 24(9):53–55

J. Zhang and P. Zhang, *Time Series Analysis Methods and Applications for Flight Data*, DOI 10.1007/978-3-662-53430-4

20. Liang J, Dun J, Sun X (2006) A combined algorithm for high-dimensional similarity search in time series database. Comput Engine 32(10):172–174
21. Dariusz W (2000) Neural network system for helicopter rotor smoothing. In: IEEE aerospace conference proceedings. Big Sky, USA, pp 271–276
22. Markus P (1996) Neural network based design for a helicopter flight mechanic model used in a heterogeneous distributed workstation architecture. In: Fuzzy systems proceedings of the 15th IEEE international conference on volume 1, New Orleans, LA, USA, pp 368–372
23. Dennice G (2003) Fault diagnosis in gas turbine engines using fuzzy logic. In: IEEE fuzzy information processing NAFPS, North America, pp 341–346
24. Vaughn ML (2003) Explaining how a multi-layer perceptron predicts helicopter airframe load spectra from continuously valued flight parameter data. In: Proceedings of the international joint conference on neural network. IEEE, pp 1059–1064
25. Marcello R (1999) Napolitano neural and fuzzy reconstructors for the virtual flight data recorder. IEEE Trans Aerosp Electron Syst 35(1):61–71
26. Wang Y, Xie S (2003) Method for quantification of aeroengine performance tendency monitoring. J Aerosp Power 18(4):549–553
27. Zhu J (2003) The developing tend of aeroengine control. Aircr Des 1:58–61
28. Sun Y, Jing B (2005) Consistent and reliable fusion of multi-sensor based on support degree. J Transcluction Technol 18(3):537–539
29. Sun Y, Zhang J, Jing B (2004) A new method to improve the distributed inspection data characteristic. Electr Measure Instrum 41(6):8–10
30. Yu C, Xu H, Huang S (2005) A relation matrix method for multisensor data fusion. Aeronaut Comput Tech 35(1):23–26
31. Hu Z, Liu X (2005) Improved consensus data fusion algorithm. J Transducer Technol 24(8):65–70
32. Duan Z, Han C, Tao T (2005) Consistent multi-sensor data fusion based on nearest statistical distance. Chin J Sci Instrum 26(5):478–481
33. Jiang H (2011) G-factor flight data pre-processing and its procedures optimization. Comput Eng 37(11):291–293
34. Zhang P, Li X, Sun Y et al (2007) Amnesic multi-sensor fusion algorithm for streaming time series. In: Proceedings of 2007 8th international conference on electronic measurement & instruments, vol IV, Cui Jianping. IEEE Press, Xi'an, China, pp 50–54
35. Box GEP (1994) Time series analysis: forecasting and control. Prentice-Hall Press, San Francisco, pp 18–43
36. Han J, Kamber M (2001) Data mining: concepts and techniques. Academic Press, San Francisco, pp 121–154
37. Yang J, Gong F (2011) Improved dynamic weighted multi-sensors data fusion algorithm. Comput Eng 37(11):97–99
38. He S (2003) Applied time series analysis. Peking University Press, Beijing, pp 47–57
39. Zeng Z (2005) Dynamic data model. Tianjin University Press, Tianjin, pp 88–90
40. Wu J, Chen Z, Li S (2007) Study on computer processing of recorded flight data. Comput Simul 24(2):18–21
41. Lv Y, Lang R (2010) Noise reduction method for flight data based on singular value decomposition. Comput Eng 36(3):260–262
42. Gong J, Zhang J, Li X (2002) A review of the development of virtual sensor and data fusion. Mech Sci Technol 21:54–57
43. Huang X (2005) Aeroengine virtual sensor based on fuzzy logic. J Nanjing Univ Aeronaut Astronaut 37(4):447–451
44. Dong J (2008) Construction of virtual sensor based on polynomial predictive filter. Chin J Sci Instrum 29(7):1408–1413
45. Zhang P, Li X (2004) RBF neural network fusion method in the application of the virtual flight parameter recorder. In: 3rd signal and information processing national conference, Ya'an, pp 326–329

46. Glibogel C, York JA (2001) Chaotic impact on science and society (trans: Yang L et al). Hunan Science and Technology Press, Changsha
47. Li T-Y, Yorke JA (1975) Period three implies chaos. Am Math Monthly 82(10):985–992
48. Agrawal R, Faloutsos C, Swami A (1993) Efficient similarity search in sequence databases. In: Proceedings of the fourth international conference on foundations of data organization and algorithms, Illinois, Chicago, pp 69–84
49. Keogh E (1997) Fast similarity search in the presence of longitudinal scaling in time series databases. In: Proceedings of the 9th international conference on tools with artificial intelligence. IEEE, Newport Beach, pp 578–584
50. Chan K, Fu AWC (1999) Efficient time series matching by wavelets. In: Proceedings of the international conference on data engineering, Sydney, Australia, pp 126–133
51. Lam SK, Wong MH (1998) A fast projection algorithm for sequence data searching. DKE 28(3):21–339
52. Chen L, Ozsu MT (2004) Multi-scale histograms for answering queries over time series data. In: Proceedings of 20th international conference on data engineering, Boston, USA, p 838
53. Jiang R, Li D (2000) Time series similarity search based on the form. J Integr Plant Biol 37(5):601–608
54. Zheng B, Xi Y, Du X (2002) Research on similarity mining in time series data sets. Control Decis 17(5):527–531
55. Wang D, Rong G (2004) Pattern distance of time series. J Zhejiang Univ (Eng Sci) 38(7):395–398
56. Li A, Qin Z (2005) Dimensionality reduction and similarity search in large time series databases. Chin J Comput 28(9):1467–1495
57. Zhang JY, Pan Q, Zhang P (2007) Similarity measuring method in time series based on slope. Pattern Recogn Artif Intell 20(2):271–274
58. Zhang P, Li X, Zhang J (2008) Included angle distance of time series and similarity search. Pattern Recogn Artif Intell 21(6):763–767
59. Dai S (2005) Data compression. Xi'an Electronic Science and Technology Press, Xi-an, pp 122–129
60. Yue F, Sun L, Wang K (2008) State-of-the-art of cluster analysis of gene expression data. Acta Automatica Sinica 34(2):113–120
61. Zhao Z, Liu F (2008) Application research of statistical monitoring index based on mahalanobis distance. Acta Automatica Sinica 34(2):493–495
62. Liu S, Jiang H (2004) A review on time series representation for similarity-based pattern search. Comput Eng Appl 27:53–59
63. Faloutsos C, Ranganathan M, Manolopoulos Y (1994) Fast subsequence mathing in time-series database. In: Proceedings of 1994 ACM SIGMOD international conference on management of data. ACM, New York, USA, pp 419–429
64. Ripley BD (1996) Pattern recognition and neural networks. Cambridge University Press, Cambridge, pp 15–35
65. Refier D, Mendelzon A (1997) Similarity-based queries for time series data. In: Proceedings of the ACM SIGMOD international conference on management of data. ACM, New York, USA, pp 13–25
66. Jin X, Lu Y, Shi C (2000) Micro-distance discovery in time series database. J Comput Res Dev 37(9):1064–1070
67. Zhang H, Cai Q (2003) Time series similar pattern matching based on wavelet transform. Chin J Comput 26(3):373–377
68. Peng H (2005) Time series similar mining based on wavelet transformation. Sichuan Univ Sci Technol 24(1):89–91
69. Qin J, Wang S, Song H (2004) Algorithm for finding similar patterns over time-series data based on wavelets and feedback. J Beijing Inst Technol 24(12):1070–1073

70. Liu B, Wang W, Shi B (2006) The tight estimation distance using wavelet. J Comput Res Dev 43(10):1732–1737
71. Guan H, Jiang Q, Wang S (2009) Pattern matching method based on point distribution for multivariate time series. J Softw 20(1):67–69
72. Burrus CS, Gopinath RA, Guo H (1998) Introduction to wavelets and wavelet transform. Prentice Hall, London, pp 5–18
73. Wu YL, Agrawal D, Abbadi AE (2000) A comparison of DFT and DWT based similarity search in time-series databases. In: Proceedings of the ninth international conference on information knowledge management CIKM, New York, USA, pp 488–495
74. Hotta K (2004) View-invariant face detection method based on local PCA cells. Adv Comput Intell Intell Infom 8(2):130–139
75. Kawagoe K, Ueda T (2002) A similarity search method of time series data with combination of Fourier and wavelet transforms. In: Proceedings ninth international symposium on temporal representation and reasoning, TIME. IEEE Computer Society, Washington DC, USA, pp 86–92
76. Korm P, Sidiropoulos N, Faloutsos C et al (1996) Fast nearest-neighbor search in medical image database. In: Proceedings of 22th international conference on very large data bases, Bombay, India, pp 215–226
77. Keogh E, Pazzzni M (1998) An enhanced representation of time series which allows fast and accurate classification. In: Clustering and relevance feedback. Proceedings of the 4th international conference on knowledge discovery and data mining. AAAI Press, pp 239–241
78. Keogh E, Smyth P (1997) A probabilistic approach to fast pattern matching in time series database. In: Proceedings of the 3rd international conference on knowledge discovery and data mining. AAAI Press, pp 20–24
79. Keogh E, Pazzzni M (1999) An indexing scheme for fast similarity search in large time series databases. In: Proceedings of the 11th international conference on scientific and statistical database management. IEEE Computer Society, Washington DC, USA, pp 56–67
80. Keogh EJ, Pazzani MJ (2000) Scaling up dynamic time warping for data mining applications. In: Proceedings of the 6th ACM SIGKDD international conference on knowledge discovery and data mining. Kruskall, Boston, pp 285–289
81. Perng C-S, Wang H, Zhang SR (2000) Landmarks—a new model for similarity-based pattern querying in time series databases. In: 16th international conference on data engineering ICDE, Los Alamitos, pp 33–42
82. Sidiropoulos ND, Bros R (1999) Mathematical programming algorithms for regression-based non-linear filtering in N-dimensional real space. IEEE Trans Signal Process 47(3):771–782
83. Gionis A, Indyk P, Motwani R (1999) Similarity search in high dimensional via hashing. In: Proceedings of the 25th international conference on very large data bases. Morgan Kaufmann Publishers Inc., San Francisco, pp 518–529
84. Agrawal R, Lin K-I, Sawhney HS et al (1995) Fast similarity search in the presence of noise, scaling, and translation in time-series database. In: Proceedings of the 21st international conference on very large data bases. Morgan Kaufmann publishers Inc., San Francisco, CA, USA, pp 490–501
85. Berndt DJ, Clifford J (1996) Finding patterns in time series: a dynamic programming approach. In: Advances in knowledge discovery and data mining. AAAI/MIT Press, pp 229–248
86. Keogh E, Pazzzni M (1998) An enhanced representation of time series which allows fast and accurate classification. In: Clustering and relevance feedback. Proceedings of the 4th international conference on knowledge discovery and data mining. AAAI Press, pp 239–241
87. Zhang P, Zhang J (2007) Similarity search in time series database based on SOFM neural network. In: The 3rd international conference on natural computation, vol 2. IEEE Computer Society, Haikou China, pp 715–718

88. Lee S, Kwon D, Lee S (2004) Minimum distance queries for time series data. J Syst Softw 69(1–2):105–113
89. Berndt DJ, Clifford J (1994) Using dynamic time warping to find patterns in time series. In: Proceedings of the KDD workshop, Seattle, WA, pp 359–370
90. Halkidi MB (2001) On clustering validation techniques. J Intell Inf Syst 17(2–3):107–145
91. Keogh EJ, Pazzani MJ (1999) Scaling up dynamic time warping to massive datasets. In: Proceedings of the 3rd European conference on principles of data mining and knowledge discovery. Springer, London, pp 1–11
92. Keogh E (2002) Exact indexing of dynamic time warping. In: Proceedings of the 28th international conference on very large data bases. VLDB Endowment, pp 406–417
93. Fu AWC, Keogh E, Yung L, Lau H et al (2008) Scaling and time warping in time series querying. Int J Very Large DataBases 17(4):899–921
94. Zhang P, Zhang J, Du J, Li X (2008) Method for similar pattern discovery in time series based on neural network. Pattern Recogn Artif Intell 21(3):401–405
95. Xiao H, Hi Y (2005) Data mining based on segmented time warping distance in time series database. J Comput Res Dev 42(1):72–78
96. Beckmann N, Kriegel HP, Schneider R (1990) The R*-tree: an efficient and robust access method for points and rectangles. In: Proceedings of the ACM-SIGMOD international conference on management of data, Atlantic City, NJ, pp 322–331
97. Jianhai Liang (2005) The intelligent approach to flight data processing based on SOFM neural network derived from principle character drill. J Projectiles Rockets Missiles Guida 25(2):182–185
98. Liang J, Du J, Sun X (2006) A WSTB-based algorithm for similarity search in time series database. Comput Eng 32(1):48–50
99. Cui B, Teng S, Cui Z (2011) Similarity search over time series data based on Walsh transform. Comput Eng 37(8):55–57
100. Zhang P, Zhang J, Li X (2004) A study of aircraft health status evaluation based on flight data trend monitor. J Air Force Eng Univ (Nat Sci Ed) 5(3):8–10
101. Yang S, Wu Y (1991) The engineering application of time series analysis. Huazhong Science University Press, Wuhan
102. Han M (2001) The challenge of data mining and stactistics. Stat Res 8:55–57
103. Wang X, Cao L (2002) GA algorithm—theory, application and software realization. Xi'an Jiaotong University Press, Xi-an, pp 77–197
104. Dai D, Xiong Y, Zhu Y (2010) Efficient algorithm for sequence similarity search based on reference indexing. J Softw 21(4):718–723
105. Song F, Gao X, Liu S (2005) Dimensionality reduction in statistical pattern recognition and low loss dimensionality reduction. Chin J Comput 28(11):1915–1922
106. Siedlecki W, Sklansky J (1988) On automatic feature selection. Int J Pattern Recogn Artif Intell 2(2):197–220
107. Wang H, Bell D, Murtagh F (1999) Axiomatic approach to feature subset selection based on relevance. IEEE Trans Pattern Anal Mach Intell 21(3):271–277
108. Narendra PM, Fukunaga K (1977) A branch and bound algorithm for feature subset selection. IEEE Trans Comput 26(9):917–922
109. Bruzzone L, Roli F, Serpico SB (1995) An extension of the Jeffreys_matusita distance to multiclass cases for feature selection. IEEE Trans Geosci Remote Sensing 33(6):1318–1321
110. Krusinska E (1988) Robust methods in discriminant analysis. Rivista di Statistica Applicada 21(3):239–253
111. Zhang J, Pan Q, Liang J (2007) A shrinking-clustering method for high dimensional data using flexible size grid. Pattern Recogn Artif Intell 20(5):716–721
112. Jiang Z, Cai Z, Wang Y (2010) Hybrid self-adaptive orthogonal genetic algorithm for solving global optimization problems. J Softw 21(6):1296–1307
113. Zhuang J, Yang Q, Du H (2010) High efficient complex system genetic algorithm. J Softw 21(11):2790–2801

114. Zhang D, Wang Z (2003) VPRS model approach to fault feature selection and diagnostic rules extraction. Acta Simulata Systematica Sinica 15(6):793–796
115. Wang L, Yu J (2005) Fault diagnosis based on discrete particle swarm optimization and support vector machine. J East China Univ Sci Technol 31(5):697–700
116. Jiang T, Li Y (2002) A dynamic identification model of aeroengine starting process based on the RBF network. J Aerosp Power 3:381–384
117. Yuan S, Du H (2005) Mechanical fault diagnosis based on artificial immune system and support vector machines. Manuf Technol Mach Tool 10:28–31
118. Angelakis C, Loukis EN, Pouliezos AD (1998) A neural network-based method for gas turbine blading fault diagnosis. ASME J Eng Gas Turbine Power 4(6):165–174
119. Han T, Yang B-S, Choi W-H (2006) Fault diagnosis system of induction motors based on neural network and genetic algorithm using stator current signals. Int J Rotating Mach 1–13
120. Shi D, Qu L (2000) The application of GA algorithm in fault feature selection. Vibr Test Diagn 20(3):171–176
121. He X, Zhu K, Wu S (2010) Thinned array synthesis based on integer coded genetic algorithm. J Electron Inf Technol 32(9):2277–2281
122. Flangn JA (1996) Self-organizing in Kohonen's SOM. Neural Netw 9:1185–1197
123. Kennedy RL (1998) Solving data mining problems through pattern recognition. Prentice Hall, Englewood Cliffs, NJ, pp 66–109
124. Wang Y (1998) Artificial intelligence principle and method. Xi'an Jiaotong University Press, Xi-an, pp 43–65
125. Huang K (1998) Expert system introduction. Southeast University Press, Nanjing, pp 22–43
126. Tang C, Qu J, Gao F (2011) Flight data criterion and its application. Comput Eng 37(10):281–283
127. Chen Y, Ni S, Huang Z (2002) Design of aeroengine life monitor system based on flight data. Aeroengine (4):12–15
128. Feng B (1992) Practical expert system. Electronic Industry Press, Beijing, pp 79–140
129. Zhang P, Wang G, Zhou D (2000) Fault diagnosis of dynamic system. Control Theory Appl 17(2):153–158
130. Zhang J (2000) The information in the fault diagnosis system. Basic Autom 7(4):6–9
131. Yang R, Chen Z (2000) Computer control system fault diagnosis. Basic Autom 7(3):35–36
132. Venk V, Yamamoto Y (1990) Process fault detection and diagnosis using neural networks-I. Comput Chem Eng 14(7):699–712
133. Zhou D, Wang G (1998) Review of fault diagnosis techniques, vol 25(1). Control and Instruments in Chemical Industry, pp 58–62
134. Zhou D, Ye Y (2000) Fault diagnosis technique and control. Tingshua University Press, Bejing, pp 66–210
135. Wen C, Hu J, Wang T (2008) Relative PCA with applications of data compression and fault diagnosis. Acta Automatica Sinica 35(9):1129–1140
136. Li G, Zhang P, Li X (2008) Sensor fault detection based on dynamic principle component analysis. J Data Acquis Process 23(3):338–341
137. Cai Z, Xu G (1996) Artificial intelligence and application. Tsinghua University Press, Beijing, pp 66–108
138. Zhang Y, Zhou H, Qin S (2010) Decentralized fault diagnosis of large-scale processes using multiblock kernel principal component analysis. Acta Automatica Sinica 36(4):593–597
139. Dong J, Wang S, Xiong F (2012) Affinity propagation clustering based on variable-similarity measure. J Electron Inf Technol 34(9):2200–2207
140. Yu H, Chen C, Zhang X (2000) Intelligent diagnosis based on neural network. Metallurgical Industry Press, Beijing, pp 78–142
141. Ankerst M, Breunig M, Kriegel H-P, Sander J (1999) OPTICS: ordering points to identify the clustering structure. In: Proceedings of 1999 ACM-SIGMOD international conference management of data, Philadelphia, PA, pp 49–60

142. Huo Z (2001) Research on fault diagnosis based on neural networks. Ind Control Comput 14(10):19–21
143. Chang Y, Zhang P (2007) Probabilistic neural network based aeroengine fault diagnosis. Microcomput Inf 23(9):177–178
144. Qiu B, Shen J (2006) Grid-based and extend-based clustering algorithm for multi-density. Control Decis 21(9):1011–1014
145. Kember G, Fowler AC (1993) A correlation function for choosing time delays in phase portrait reconstructions. Phys Lett A (2):72–80
146. Fu Z, Cheng W, Xu C (2009) LS-SVM-based method for modal parameter identification. Acta Aeronautica Et Astronautica Sinica 30(11):2087–2092
147. Ma S, Wang T, Tang S et al (2003) A fast clustering algorithm based on reference and density. J Softw 14(6):1089–1095
148. Yue Q, Feng S (2009) The statistical analyses for computational performance of the genetic algorithms. Chin J Comput (12):2389–2392
149. Ye M, Cheng Y (2004) Hierarchical clustering algorithm based on distribution model. J Univ Electron Sci Technol China 33(2):171–174
150. Huang J, Li S, Cai Y (2005) An approach to building rough data model through supervised fuzzy clustering. J Softw 16(5):744–753
151. Zhang P, Li X (2007) Design and realization of the remote fault diagnosis system for the airplane based on Matlab Web Server. Microcomput Inf 23(8–1):190–191
152. Gao J, Xu Y, Li X (2007) Weighted-median based distributed fault detection for wireless sensor networks. J Softw 18(5):1209–1217
153. Kugiumtzis D (1996) State space reconstruction parameters in the analysis of chaotic time series—the role of the time window length. Physica D 1:13–28
154. Lin J, Wang Y, Huang Z et al (1999) Voice signal time delay in phase space reconstruction. Complex Self-correlation Algorithm (3):220–225
155. Zhang X, Wang H (2010) Condition time series prediction using least squares support vector machine with adaptive embedding dimension. Acta Aeronautica Et Astronautica Sinica 31(12):2309–2314
156. Li X, Wang Z, Lv T (2010) A novel incremental clustering algorithm based on chaos and immune response. Acta Automatica Sinica 36(2):208–214
157. Kim HS, Eykholt RJ, Salas D (1999) Nonlinear dynamics, delay times, and embedding windows. Physica: D 1:48–60
158. Nicolis C (1987) Climate predictability and dynamical system. In: Proceeding of the NATO advanced system study institute on irreversible phenomena and dynamical system analysis in geosciences, Crete-Greece (1985). NATO ASI Ser. C. D. Reidel, Hingham, MA, pp 321–354
159. Taylor JH, Sharif SS (2007) Chaos in nonlinear dynamic systems: helicopter vibration mechanisms. In: Mediterranean conference on control and automation, Athens, Greece
160. Liu B, Zhang J, Zhang P (2010) Similarity search method in time series based on curvature distance. J Electron Inf Technol 32(3):509–514
161. Feng H, Hu Z (2001) Data mining technology in the application of the UNIX system performance analysis. J Northwest Polytechnical Univ (3):88
162. Tang G (1998) Investigations on selecting performance-monitoring parameters of aeroengine. J Propul Technol (2):38–53
163. Ma J, Liu Q, Ma Y (2009) Detection of chaotic vibration signals of aero-engines with blind separation. Acta Aeronautica Et Astronautica Sinica 30(1):81–85
164. Hu J (2003) Research about trend of a certain type of engine performance monitoring and fault diagnosis system. The Air Force Engineering University, Xi-an
165. Zhang P, Chang Y (2007) Study on dynamic process clustering based on artificial neural network. Comput Eng Appl 43(27):72–73
166. Vitányi PMB (2011) Information distance in multiples. IEEE Trans Inf Theory 57(4): 2451–2456

167. Kim JH, Lee SH, Wang H (2009) Similarity measure application to fault detection of flight system. J Cent South Univ Technol 16:789–793
168. Vlachos M, Kollios G, Gunopulos D (2002) Discovering similar multidimensional trajectories. In: Proceedings of the 18th international conference on data engineering, San Jose, USA, pp 673–684
169. Li Z, Zhang F, Li K (2011) DTW based pattern matching method for multivariate time series. Control Decis 26(4):565–570
170. Zhang Y, Guan W (2009) Calculation Lyapunov exponent spectrum of time series based on least-squared support vector machine. Comput Eng Appl 45(31):196–199
171. Lu F, Huang J (2009) Engine component performance prognostics based on decision fusion. Acta Aeronautica Et Astronautica Sinica 30(10):1795–1800
172. Yao M, Qi D, Zhao G (2002) On control of discrete-time chaotic systems based on Lyapunov exponents. Control Decis 17(2):171–174
173. Zhang J, Pan Q, Song J (2007) A monitoring and forecasting method of airplane status based on WLS-SVM. J Air Force Eng Univ (Nat Sci Ed) 8(6):1–4
174. Liu Q, Qin Z, Chen W et al (2011) Zero-norm penalized feature selection support vector machine. Acta Automatica Sinica 37(2):252–256
175. Zhang P, Li G, Li X (2006) Aviation automation information processing system based on flight data. Aviat Maintenance Eng 6:37–39.
176. Singhal A, Seborg DE (2001) Matching patterns from historical data using PCA and distance similarity factors. In: Proceedings of the 2001 American control conference, vol 2, Arlington, pp 1759–1764
177. Sun T (2003) Research on the application of flight data. Nanjing University of Aeronautics & Astronautics, Nanjing

Printed in the United States
By Bookmasters